Josef X. Brunner Gunther Wolff

Pulmonary Function Indices in Critical Care Patients

Springer-Verlag
Berlin Heidelberg New York
London Paris Tokyo

Josef X. Brunner, Dr. Sc. techn.

Address during preparation of this book:
Research Assistant of Clinical Physiology
University of Basel
Kantonsspital
CH-4031 Basel
Switzerland

Current address:
Department of Anesthesiology
Division of Biomedical Engineering
3 C 444 University of Utah Medical Center
Salt Lake City, Utah 84132
USA

Gunther Wolff, M. D., FMH Surgery
Professor for Surgery
and Intensive Care Medicine
Head of Clinical Physiology

Address:
Clinic for Thoracic and Cardiac Surgery
Department of Surgery
University of Basel
Kantonsspital
CH-4031 Basel
Switzerland

ISBN-13: 978-3-540-18432-4 e-ISBN-13: 978-3-642-73040-5
DOI: 10.1007/978-3-642-73040-5

Library of Congress Cataloging-in-Publication Data. Brunner, Josef X., 1956-. Pulmonary function indices in critical care patients. Bibliography: p. Includes index. 1. Pulmonary function tests. 2. Critical care medicine. I. Wolff, Gunther, 1932-. II. Title.
RC 734.P84B78 1988 616.2'4'028 88-3161

Foreword

Respiration is a unique topic among various subdisciplines of physiology. Physiologists and clinicians are now able to communicate quantitative functional properties of lung mechanics and gas exchange in the language of the engineer, physicist and mathematician. This is largely due to intensive and stimulating work during the last decades of brilliant minds in a handful of excellent schools in the international family of physiologists. Among these founders of respiratory physiology are a number of clinicians, and they have taken significant part both in shaping the theoretical knowledge to clinical applicability and developing technical devices for diagnosis and therapy in pulmonology. However, the theory behind the evaluation of measurements, and their interpretation in terms of clinical function tests, is so confusingly complex that the ordinary physician, not specifically trained in respiratory physiology, finds himself unable to critically apply these techniques. We, therefore, need descriptions of respiratory physiology and of its clinical application presented in the language of the clinician.

And that is what this book is meant to be. Written by an expert in electrical and biomedical engineering, and by an expert in intensive care medicine, this text constitutes an "operational manual" of clinical respiratory physiology. It does not intend to be another textbook of basic respiratory physiology or pathophysiology. This book not only addresses practical clinicians, particularly those of intensive care medicine, by describing the essentials of clinically relevant respiratory knowledge. But it also addresses the engineer, by describing the essentials of biomedical engineering relevant to lung function analysis. The reliability and the elegance of the measuring techniques described in this book are to be commended. And what makes the authors eligible for such a task is that they have plunged deep into the understanding of the physiological basis and have, at the same time, kept their view as practical clinicians, since they maintain an everyday responsibility for the intensive care ward.

Hence, the book may be profitable read by clinicians, respiratory physiologists and engineers, particularly those with interests in intensive care medicine and pulmonology, who wish to learn details concerning both the theoretical background and the practical aspects of clinical function tests.

I trust that this book will find friendly acceptance by the colleagues in the field.

October 1987

Prof. Peter Scheid

Head of the Institute of Physiology
Ruhr-Universität
Bochum
West-Germany

Preface

This book started out 10 years ago with the question: How do I measure functional residual capacity in the intensive care patient? The following 5 years were filled with disappointment, mainly caused by the expensive measuring equipment. We were not able to perform measurements without running into serious limitations, and the algorithms were "buggy". The measurements turned out to be inaccurate and, worst of all, they could not be verified. The only solution to these problems seemed to be developing a custommade system. In the end, this decision proved to be the right one.

The task required an additional member in the team: an engineer. The first problem was communication. It turned out that even though they spoke the same language, the words did not mean the same for the engineer as they did for the physician. Almost completely opposite were the opinions as to what kinds of problems a computer can solve. Soon, another problem was discovered: the scatter of information. Although many of the techniques needed had already been developed, in physiology as well as in engineering, there was no common source of information. This book, which is the enhanced and translated version of a PhD thesis (Brunner 1985), attempts to fill the gap. We tried to cover most of the terminology used in both, medicine and engineering, with respect to lung function analysis. The prime goal is to enable efficient communication between physician and engineer in order to evaluate commercially available equipment or to design new equipment.

There is no claim for completeness regarding engineering or pulmonary physiology. We strongly recommend that the original literature be consulted. In particular, we encourage the reader to familiarize himself with the concepts developed by Fenn, Rahn, and Farhi in Rochester, which are not dealt with in detail in this book (V′/Q′ diagram, see Chapter 2.4.3). As an excellent source of information we also recommend the "Handbook of Physiology" published by the American Physiological Society.

The success of the close (and pleasant) collaboration of physician and engineer extended the original project far beyond the initial plan to measure functional residual capacity. This book is one of the results. Its major purpose is to save interested researchers the "first five unpleasant years".

In Chapter 1 the properties of a clinically important pulmonary function index are presented. The particular modes of application of the measuring method in the clinic and in seriously ill patients are discussed. In Chapter 2 the basic principles of model development for spontaneous respiration and mechanical ventilation are discussed. In Chapter 3 we describe our solutions for the measuring technique and analysis. We present three cases to document the efficiency of the entire measuring system. The next 3 chapters deal with applications. In Chapter 4 lung function indices of a standard situation are presented. In Chapter 5 the effects of isolated changes in individual variables of the breathing pattern are described through a comparison of various steady states (paired differences). Chapter 6 deals with the possibilities and limits of time resolution, i. e. the breath by breath analysis. Intermittent mandatory ventilation (IMV) is assessed by separate analysis of spontaneous breaths and mandatory breaths.

Appendices are included, some of which are designed to give the physicists/technologists of the team an idea of the anatomy and physiology of the lungs, others should help the physician of the team to understand the physics and mathematics involved. A further appendix contains all of the formulas used in the program package. Finally, a last appendix presents in tables all of the results of the investigations found in Chapters 4, 5, and 6.

It is our hope that this book will help to make pulmonary function testing in the critically ill patient more reliable, precise, rational, critical, and less expensive. Our goal should be to treat our patients with better differentiation and to deepen our understanding of functional disturbances. For this purpose we have made our software with documentation available to the interested reader. In return, our only request (apart from copying costs) is feedback, in the hope that such a widespread exchange of experience and information will be a valuable contribution to the fascinating discussion of pulmonary function in the critically ill patient.

January 1988

Josef X. Brunner and *Gunther Wolff*

Salt Lake City, Utah, USA and
Basel, Switzerland

Contents

Acknowledgements

This work was made possible by the exchange of ideas and extensive discussions with many of our colleagues in the scientific community. Most of these discussions took place not at large congresses, but rather in the laboratories. For their interest, support, and hospitality we are most grateful to Professor Johannes Piiper (Göttingen, W. Germany), Professor Peter Scheid (Bochum, W. Germany), Professor Gordon Cumming (Midhurst, GB), Dr. Göran Hedenstierna (Stockholm, Sweden), Professor Emilio Agostoni (Milano, Italy), and Professor Peter Suter (Geneva, Switzerland).

Interdisciplinary research is impossible without open communication between medical doctors, physiologists, and engineers. In this respect we gratefully acknowledge the invaluable discussions with Dr. Manuel Paiva (Brussels, Belgium), Dr. Chris Mills (Midhurst, GB), Professor David Dennison (London, GB), and Professor Max Anliker (Zürich, Switzerland).

Our research was fully supported by the Director of the Clinic for Cardiac and Thoracic Surgery of the University of Basel, Switzerland, Professor Erich Grädel. We are most grateful for the research opportunities he provided us with, as well as his personal support. Our thanks also go to the Head of the Department of Surgery of the University of Basel, Professor Felix Harder, for his valuable support.

Each improvement of the software was tested in our cardiac surgery intensive care unit, which interfered immensely with the work on the ward. We want to thank our nurses and their staff for their patience and cooperation. Their critical interest as users of the measuring system proved to be invaluable. Many of the hardware improvements were carried out with the hospital technicians, in particular, Werner Nyffeler. We are indebted to them for their kind assistance and constructive imagination.

The original manuscript was written in German. The translation into English proved to be an unexpectedly difficult task. Virginia Sonntag-O'Brien deserves special commendation for the translation of most parts of the book. We are further indebted to Dr. Robert Loeb and Dr. Tom East for their scientific critique and help with the English. We wish to give special thanks to Dr. Toni Graf-Baumann from Springer-Verlag, who encouraged us to write this book and whose advice and suggestions on how to improve it didactically were a major asset.

We have not forgotten that our work was only possible thanks to the financial support of many institutions. We therefore wish to express our appreciation to those who showed their confidence in our project. We were supported by the Swiss National Science Foundation, the Roche Research Foundation, the Jubilee Foundation of

Ciby-Geigy, the Emilia Guggenheim-Schnurr Foundation, the Science Foundation of the Basel Chamber of Commerce, the Sandoz Research Foundation, and the Science Foundation of the Kantonsspital-Basel. Last but not least we would like to thank Drägerwerk and Dr. Eric Hecker (Lübeck, W. Germany). He provided the opportunities for international researchers to get together at productive workshops in our department. We sincerely appreciate his interest in the physiological aspects as well as his critical and stimulating contributions to the discussions.

List of Abbreviations

General symbols

V	volume
V′	volume flow
Q′	blood flow
F	gas concentration in dry gas
P	pressure or partial pressure
C	content

General Subscripts

A	involving the alveoli
a	arterial
c	involving the lung capillaries
I	inspiratory
$\bar{v}$	mixed venous
E	expiratory

Abbreviations

ABG	arterial blood gas analysis
ADU	A/D Unit, 1/4095 of ± 5 volt
alv. eff.-CO_2	efficiency of the CO_2 elimination
APV	accessible pulmonary volume
ARI	acute respiratory insufficiency
ARDS	Adult Respiratory Distress Syndrome, acute progressive lung failure in the adult (accute, rapidly progressing parenchymatous disturbance of the lungs with very severe ARI)
$a\bar{v}D_{O_2}$	arterio-venous oxygen content difference
b. h.	body height
BTPS	body temperature & pressure of water vapor saturated gas
b. w.	body weight
C_{CO_2}	CO_2 content in blood
C_{O_2}	O_2 content in blood
C_{eff}	effective compliance
CI	cardiac index
C_L	volume distensibility of the lung

CO	cardiac output
C_{tot}	volume distensibility of the entire respiratory system
CV	variation coefficient
C_W	volume distensibility of chest wall and diaphragm
E	expiration
ECC	extracorporeal circulation (heart-lung machine)
EIP	end-inspiratory pause
$F_{CO_2a\text{-}E}$	difference between arterial and expired (end-tidal) CO_2
$F_{CO_2\text{-}et}$	maximum CO_2 concentration in the expirate
$F_{\bar{E}CO_2}$	mixed expiratory CO_2 concentration
F_{IO_2}	inspiratory O_2 concentration
$F^{\circ}_{N_2}{}^{\circ}$	initial nitrogen concentration
FRC	functional residual capacity
I:E	ratio of inspiration time to expiration time
LAP	left atrial (blood) pressure
mBP	mean (arterial) blood pressure
MS	mass spectrometer
P_A	pressure in the alveoli
P_{CO_2}	partial pressure of CO_2
P_{O_2}	partial pressure of O_2
PAP	pulmonary artery (blood) pressure
P_{awo}	pressure at the airway opening, i.e., at the mouth, at outer end of tube
P_{awoEI}	end-inspiratory P_{awo}
$P_{awoI}min$	minimal airway pressure during inspiration
$P_{awo}max$	maximum P_{awo}
$P_{\bar{E}CO_2}$	mixed expiratory partial pressure of CO_2
$P_{\bar{E}O_2}$	mixed expiratory partial pressure of O_2
PEEP	positive end-expiratory pressure ventilator setting
PIE_{CO_2}	approximative value for serial dead space from the CO_2 diagram
PIE_{N_2}	approximative value for serial dead space from the N_2 diagram
P_{es}	pressure in the esophagus
P_{pl}	pressure in the pleura
PT	pneumotachograph
P/V	pressure volume (P/V) diagram
Q'_S/Q'_T	venous admixture (shunt fraction)
R, RQ	respiratory quotient
RAP	right atrial blood pressure
R_{AW}	airway resistance
R_L	viscous resistance of lung tissue
RR	respiratory rate, breathing or ventilation frequency (i.e. both)
RRN	relative remaining nitrogen
R_{ti}	viscous resistance of muscle or connective tissue
RV	residual volume
rF	relative humidity
SAP	O_2 saturation of a pulmonary artery blood sample = mixed venous O_2 saturation
SD	standard deviation
SEM	standard error of the mean
SI	stationary interface
$slope_{CO_2}$	slope of the alveolar plateau in CO_2 diagram
S_{O_2}	O_2 saturation
STPD	standard temperature & pressure of dry gas
$S_{\bar{v}O_2}$	mixed venous O_2 saturation
T	temperature
T_I, T_E	inspiratory and expiratory time
TLC	total lung capacity
V'_A/Q'	ventilation perfusion ratio
VC	vital capacity

V'_{CO_2}	CO_2 elimination per minute
V_D/V_T	physiological dead space ratio
V_d/V_t	Bohr dead space ratio
V_{dAW}	airway dead space, other term for series dead space
V_{dS}	series dead space, other term for V_{dAW}
V_{ext}	external dead space (see Fig. 2.7)
V'_I	inspiration flow (rectangular flow)
V_L	lung volume
$V_{L\text{-}EI}$	end-inspiratory lung volume
V'_{O_2}	O_2 uptake per minute
V_T	tidal volume
WO. eff.	washout efficiency
ZEEP	zero end-expiratory pressure (i. e. ventilation without PEEP)

Conventions

The pressure units used are those that are commonly used in daily clinical practice, i. e.
mmHg for blood pressure
mmHg and (kPa) for partial pressures of O_2, CO_2, and other gases
cmH_2O for ventilatory pressure
vol% in STPD for partial pressure in gases

1. Evaluation of Pulmonary Function in the intensive Care Patient

1.1. The Clinically Important Pulmonary Function Index

1.1.1. Evaluation and Monitoring of Circulatory Functions

Beat-to-beat monitoring of electrocardiograms, arterial and central-venous pressure, and even pulmonary arterial pressure (to name just a few methods) have become absolutely routine in intensive care units and in operating theaters. The circulation monitoring devices currently being marketed by the medicalelectronics industry are so sophisticated that problems associated with calibration, zero points, or stability are just about passé. Measuring methods are designed in such a way that for the nursing staff who uses them they are simply child's play. Readings are shown at bedside on digital and graphic displays, on line and in real-time. And what is more, the pragmatic interpretation of these readings by the clinician is coming closer and closer to the interpretation made in physiological models.

Thanks to this development, the modern treatment of severe circulatory disturbances has become a blessing for the patient and a source of fascination for the physiologically interested therapy team. Our younger colleagues have already begun to take this meshing of hemodynamic physiology, diagnostics, therapy, and success for granted. It is easy, however, to overlook the fact that the clinical-physiological treatment concept so successfully applied in hemodynamics today is the result of years of development, involving extensive bedside observation of the clinical course, comparative bedside observations of changes in the values measured, and the initially only tentative interpretation of developments on the basis of physiological models.

1.1.2. Evaluation and Monitoring of Pulmonary Functions

What about the clinical-physiological treatment concept seriously ill patients with respiratory insufficiency are offered at bedside? What up-to-date information does the field of respiratory therapy have at its disposal? How can course trends in pulmonary disease be evaluated on the basis of individual critically ill patients?

Say, perhaps, we keep an hourly record of respiratory rate and auscultate the patient several times a day, even under unfavorable conditions. And several times a day we take arterial blood samples, although we often have no accurate figures on inspiratory oxygen concentration. Then we let a non-standardized time under non-standardized conditions go by before the samples are analyzed, and go on to test them using ultra-sophisticated analytical techniques and to interpret the results with scrupulous accuracy. We take a daily antero-posterior chest x-ray with the patient lying flat or slumped over the edge of the bed and stare fascinated at a picture, the quality of which would be rejected outright were it a chest x-ray produced in the outpatient department. In cases where the patient is being mechanically ventilated, the machine is – in addition – monitored breath-by-breath, and the respiratory frequency, peak airway pressure (measured haphazardly somewhere in the ventilation system), and the gas volume exhaled per breath are roughly plotted. An acoustic alarm goes off when a ventilation tube becomes disconnected, when the oxygen concentration falls below the selected limit, when peak airway pressure exceeds the limit set, and, in more elaborate machines, when the gas volume exhaled per minute does not reach the selected limit.

But what about the patient? Due to this scanty amount of information, the current functional state of the lung is interpreted according to a concept which we, faithful readers of the leading journals, are completely familiar with. A concept that (perhaps) had been elaborated on using "multiple inert gas elimination" or "computed tomography" or some "xzx – isotope imaging process" under heavenly examination conditions only a few months before in one of the world's most prestigious medical centers in a select group of (perhaps) 20 patients, who (perhaps) were in the same condition. Concepts such as these will have been painstakingly developed in that specific patient group with an investment of brains, manpower, equipment, and time that one generally cannot afford under conventional therapeutic conditions. The pathophysiological state of our patient's lungs is then labelled on the basis of this remote concept and our patient is treated according to this label by analogy with that remote patient group. The success of the therapy, and hence the effectiveness of the selected therapy, is then measured in our patient using the inadequate criteria available to us in routine daily practice.

Over the last few years, various measuring systems for the continuous registration of pulmonary functions have been developed. Thus, it is all the more surprising that none of the approaches to the continuous analysis of pulmonary function has become accepted in practice. Obviously, the advantages of a continuous stream of quantitative information have not outweighed the disadvantages of the disturbances these measurements cause. There might be two reasons for this: first, applying the measuring system in the clinic was too complicated and disrupted the primarily therapy-oriented organization, or the results were unreliable and not precise enough. These reasons are of a technical nature. Secondly, the measured values obtained with the system were not able to influence the therapeutic decisions, that is, they were not "clinically important." This reason has more to do with the difficulty (or impossibility) of interpreting the measured values. Let us discuss these interpretation difficulties more closely.

1.1.3. "Simple" Measured Values

Every system yields "simple" measured values which of course could also be obtained with less complex devices. These include single breath values such as tidal volume, respiratory frequency, maximum or end-inspiratory pressure in the respiratory system, and end-expiratory carbon dioxide concentration. Values measured over a certain period (e.g. for several minutes) and then averaged out for one minute, such as minute volume, O_2 uptake, and CO_2 elimination, are also "simple" measured values. Such values can serve as a semi-quantitative description of the condition of the entire organism when an actual measured value is compared with the reference values normally encountered in patients with similar anthropometric data and under similar conditions. While "simple" measured values allow changes to be detected as they occur, they do not suffice to describe the function of the lung.

1.1.4. Physiological Models and Pulmonary Function Indices

When using certain mechanisms to explain the functions of the lung, we mentally replace these with a model. The more accurately the function of the lungs (output of the model) can be predicted, the more useful the model. Furthermore, the model must be consistent with all of the features of the lung devised by other methods (e.g. anatomical findings). Applying the model allows a description of the lungs in the form of a number: a so-called pulmonary function index. It thus follows that the numerical value of a specific pulmonary function index is only valid with reference to a specific model. Moreover, the usefulness of a pulmonary function index depends not only on the accuracy of the measurements performed in the patient, but also on the value of the underlying lung model. For the engineer, the model determines the list of physical parameters to be measured on the patient and the accuracy with which they have to be measured. For the clinician, the model shows the limits within which the measured pulmonary function index may be interpreted. Discussion of the models is no mere academic diversion; it is extremely important and must be conducted thoroughly.

1.1.5. The Clinically Important Pulmonary Function Index

We want to close this section by listing the requirements a clinically important pulmonary index should meet:

1. A clinically relevant pulmonary index should be based on a non-contradictory, valid physiological model.
2. The primary data needed to calculate the index should be measured non-invasively and by simple means.
3. It should be possible to measure continuously breath by breath or at least repeatedly at short intervals and as often as desired.
4. Derivation of the index (investigations in the patient, measurement of the primary data, calculation of the index) should require very little time (on line, real time).
5. The entire sequence (investigation, measurement, calculation, presentation of the results) to determine an index (such as V or FRC) should have been tested on a physical model.
6. Normal values should have been measured in normal subjects and standard values in patients in clinically defined situations or with clinically defined syndromes. Reproducibility should have been tested.
7. The index should be able to influence important therapeutic decisions.

For the sake of completeness, it should be added that one could accidentally discover a quantifiable feature which is not based on a physiological model, but which proves to be a successful indication for a particular therapy. In this case, and although condition 1 is not fulfilled, one would still have to speak of a clinically important index.

1.2. What Clinicians Expect of Transducers and Data Processing

A measuring method is usually introduced primarily through the initiative of a very small group of interested doctors (or even of one individual). The round-the-clock conditions prevailing in intensive medicine result in other colleagues also using the method on a secondary basis. If the measuring system proves suitable for monitoring purposes, it will also soon be used by the intensive care nursing staff, who generally are highly skilled in using technical instruments. While the prime users will be constantly aware of the weaknesses of the algorithms used (model, measurement, software) the secondary users will tend to take the digitally displayed data at face value under all circumstances. They will be content to use them without bothering to find out whether the actual conditions of the examination justify the use of these algorithms. The primary users will also bear in mind the delicacy and fragility of certain modules (hardware) when handling the instruments. The secondary users will treat these instruments from the very beginning as if they were as indestructible as tractors. Consequently, the team as a whole (primary and secondary users) will only accept a device if it produces plausible results even with incorrect handling, if it doesn't get in anyone's way, and if it promises to carry on working unattended even during cardio-pulmonary resuscitation procedures. Finally, it has to survive the toughest of all consumer tests, namely, cleaning and sterilization (or disinfection), which usually take place behind the hospital scenes. The following is an attempt to derive some of the features which should be expected from such instruments on the basis of the requirements listed above.

1.2.1. The Chain of Instruments: Data Flow

Measuring instruments, patients, and nursing staff are sensitive and can get in each other's way. If it is desirable to minimize these mutual disturbances in the intensive care unit, no instrument should impede access to the patient. In other words, most of the instruments should be set up away from the patient. They should only be linked to the patient with simple and totally reliable wiring. The processed information should, in turn, be made available directly at the patient's bed. The result is a chain of components, which has the task of transporting the data from the patient to the display.

1.2.2. Measuring Head and Measuring Instruments

The assessment of various pulmonary indices requires the measurement of gas composition (O_2, CO_2, N_2 tracer gas), gas flow, and gas pressure at the mouth.

Integrating these measurements in one "measuring head" offers many advantages. First of all, such an arrangement makes it possible to link the patient to the measuring system with one single manipulation. Secondly, the dead space of the instrumentation can be minimized. Thirdly, errors due to signals measured at different measuring sites can be avoided. As a consequence, corrections based on assumptions that might not always hold true later in the investigation are not necessary. Measuring all of the signals at the same place – namely, at the mouth – a priori avoids such errors. The measuring head, which is linked directly to the patient, has to fulfill two important requirements: it must be small and light so as not to disturb the patient and his care, and it should be easy to sterilize (or disinfect) before use for another patient.

For technological reasons, it is not possible to combine all the sensors in the measuring head. The sensors are rather connected to the measuring head by means of disposable catheters. While this approach is perfect in terms of hygiene, it causes problems with transmitting information.

The continuous determination of gas composition at a distance from the measuring site presupposes the transportation of a continuous gas sample from measuring site to measuring instrument, i. e., there is a time lag during gas transport from measuring site to measuring instrument. This causes no problems if all lag times are equal, but severe problems if they are not. If complex quantities are to be calculated from the different signals, the difference in the individual delay times must be exactly compensated for.

To reduce measurement errors, the measuring instruments should be close to the patient. On the other hand, the instruments should be far away from the patient in order not to disturb him or her and not to impede nursing care. It is therefore always necessary to find a compromise for the best location for the instruments: in a location where it is easier to take into account the technical requirements posed by the equipment.

The instruments are generally operated by doctors and nurses and moved and cleaned by the auxiliary staff. Preference is given to instruments that are immediately ready to use at any time without calibration and warm up time and which have baselines that do not drift and, in addition, the instruments must be absolutely shockproof and it must be possible to wipe them down with a damp cloth.

1.2.3. Data Processing

Data processing must be structured to allow for subsequent changes. Changes must be possible at the level of correction, calibration, and synchronization of raw data, as well as at the level of calculating lung function indices. Computer programs should consequently be written in a higher level language.

1.2.4. Result-output

For clinical purposes (monitoring and therapy control), the smallest number of indices should, in principle, be presented in a clear and easy-to-read manner. Values necessary in setting the machine or for monitoring the patient and machine should be represented breath-by-breath. Here, scrolling the last 5 (for instance) breaths has proved to be an effective display method, since it permits rapid plausibility control and is a good didactic aid. An equally favorable effect is achieved by the breath-by-breath graphic representation of airway pressure, gas flow, or CO_2 concentration versus time or versus exhaled volume. Values that require steady state conditions should preferably be represented in the form of mean values. It should be possible to continuously register the most important data and indices breath-by-breath.

A sizable problem mainly affecting the breathing mechanics – and one which to date has not been satisfactorily solved – is recognizing spontaneous breaths (SB) and mandatory breaths (MB), such as with intermittent mandatory ventilation. Thus, for example, certain pressure values in MB have a different significance than the same ones in SB, such as P_{awoEI} or $P_{awoI}min$ and in SB, the total airway resistance and the compliance of the total respiratory system do not exist, thus rendering the values deduced invalid. Relationships are still more complex when it comes to the new mixed forms of breathing and ventilation, such as assisted spontaneous breathing, where the individual breath begins with a spontaneous breath and then may change into positive pressure ventilation after perhaps half the V_T. Errors in interpreting these values can only be avoided if one has a breath-by-breath breakdown into patient activity and machine activity. There is a lack of clear definitions in the case of these mixed forms. Consequently, one cannot always be sure whether the measuring system is producing results which have been determined under inadmissable conditions, i.e., indices which are not regarded as "false" but rather as "invalid". Only investigators who observe the patient, supervise the measurements, and receive the data, can prevent the confusion that would result from invalid data erroneously being taken as results.

It follows that those in charge are also responsible for making sure that everyone receiving or further processing the results is able to check independently whether the actual investigation conditions warrant referring to the "data" generated by the system as "results". It is therefore essential that all of the collaborators be acquainted with the fundamentals of the indices, i.e. with the model concept.

1.3. A Measuring System for Clinical Research

In this book we are introducing a measuring system which has been developed to meet the requirements for clinical research described above. The measuring system first of all contains several components that present a new principle: the reliable recognition of inspiration and expiration, the dynamic, automatic correction of the delay time of the signals of the gas concentrations measured with the mass spectrometer, the estimation of the anatomical dead space or series dead space (V_{ds} as pre-interface expirate (PIE)), the estimation of the slope of Phase III in a single breath diagram, and the description of the gas mixing & distribution with washout efficiency. Secondly, these new developments and the interrelationships already known have been applied with the greatest possible consistency. In doing so, it has been possible to create not only a very reliable and accurate measuring system, but also one which is particularly versatile and easy to use. The following gives an overview of the measuring system.

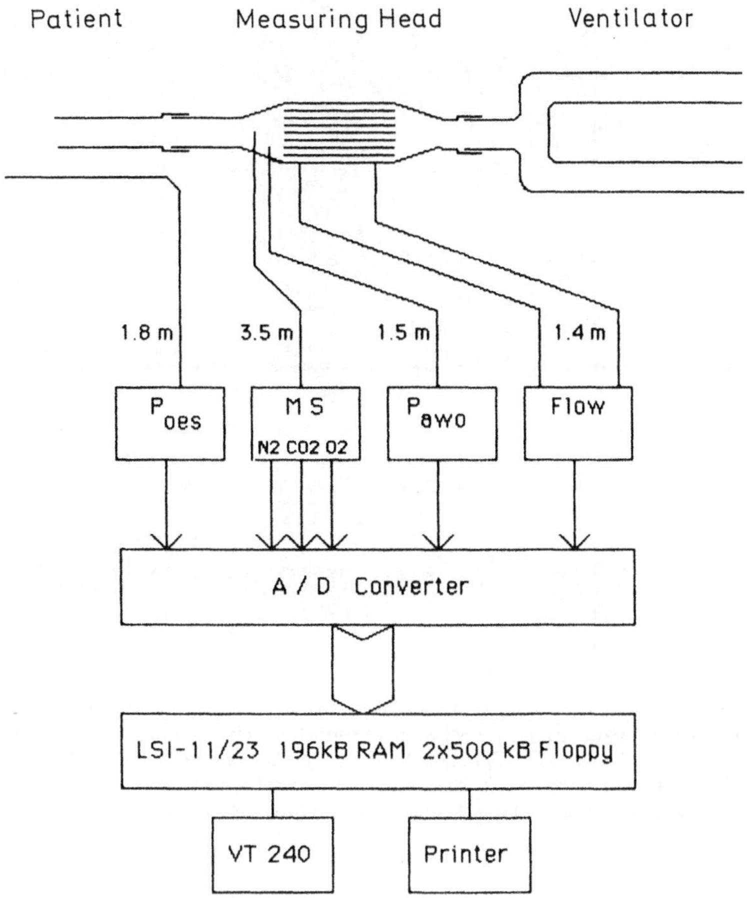

Fig. 1.1. Diagram of the measuring system. The measuring head is positioned between the patient and the ventilator. The length of the connecting tubes to the pressure transducers (P_{es} and P_{awo}) and the pressure difference transducer (flow) as well as the length of the capillary leading to the mass spectrometer (MS) are given because they determine the bandwith of the Measurements (see Chapter 3)

1.3.1. Obtaining Primary Data from Measurements at the Airway Opening

Primary data are obtained from gas-dynamic and chemical analyses of the inspiratory and expiratory gas flow. Fig. 1.1 shows a block diagram of the measuring system. In the case of ventilated patients, the usual connecting hose between patient (endotracheal tube) and ventilator is replaced with the measuring head. In spontaneously breathing patients (or test persons), the airways are connected to the measuring head by means of a mouth piece (nose clip needed).

Pneumotachograph (flow measurement), pressure measuring hoses, and the capillary leading to the mass spectrometer (see Chapter 3 for a detailed description)

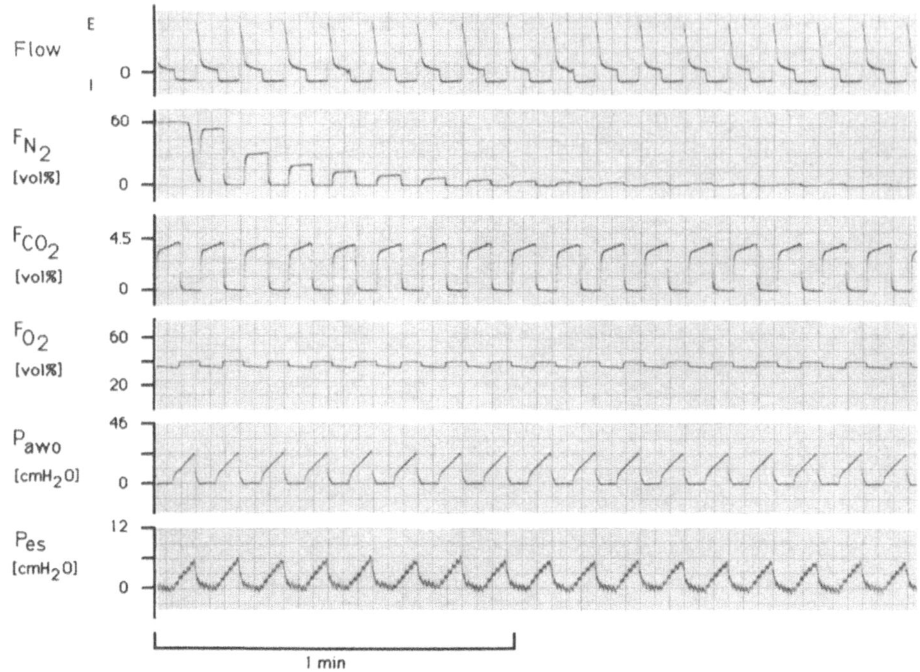

Fig. 1.2. Time course of the primary signals measured in ventilated patient shown using an example of a N_2 washout (see 2.3). Gas flow (V′) in the direction of inspiration (I) and expiration (E). The partial pressures of the gases O_2, CO_2, and N_2 are relative to the pressure of the dry gas (fractions in vol%). The concentration of argon is the difference between the sum of F_{N2}, F_{CO_2}, and F_{O_2} and 100 vol%. The esophageal pressure (P_{es}) is superimposed with cardiogenic oscillations

form a compact assembly: the measuring head. It measures the time course of the gas flow (V′), gas composition N_2, CO_2, O_2 and argon content), and airway pressure (P_{awo}) shown in Fig. 1.2. An additional probe measures the pressure in the esophagus (P_{es}).

These parameters and the time courses of all of the signals measured are used to calculate pulmonary function indices.

1.3.2. Transducers

The instruments described in Chapter 3 for measuring gas flow and pressures fulfill all of the requirements. There is a chance of the Fleisch pneumotachograph being overdriven during the short expiratory flow peak of mechanically ventilated patients. To prevent this from going unnoticed and hence uncorrected, the expiratory flow peak is displayed breath-by-breath.

The mass spectrometer used to measure gas concentrations has not only advantages but also serious disadvantages. The instrument cannot be set up directly at the bedside. It is too large, too loud, and produces too much heat. Furthermore, its performance is affected if it is moved to investigate a patient. The precision of our measurements rose as soon as the mass spectrometer was placed in a fixed location. Repairs also became less frequent. To be able to set up the mass spectrometer outside the immediate nursing area, we take the gas sample via a 350 cm long capillary from the measuring head to the measuring instrument, and solve the ensuing delay problems at the software level. Although theoretically we are able to measure at three beds with this arrangement, the measuring system is not transportable and cannot be set up at will at any bed.

The accuracy of the concentration measurements is adequate, however, the response time of the mass spectrometer (80 ms) has reached the highest tolerable limit. When mechanical ventilation is used, we are probably measuring close to the irreversible loss-of-information level (see 3.1.3ff.).

1.3.3. Preprocessing and Storage of Data

Correction, calibration, and synchronization of the signals (described in detail in Chapter 3) permits sufficiently accurate computation of the measured values and estimated indices.

All data are stored unprocessed as acquired from the A to D converter (see Fig. 1.3). Storage is done on floppy disks, which take 10 minutes of raw data each. Subsequently (off line), the stored raw data can be analyzed using more complicated programs with longer computing times or even with subsequently developed programs. The selection of the storage medium (disk, tape, etc.) of course depends on the local and financial circumstances. From the technical standpoint, the laser disk seems to be a good solution. An optical storage disk (DRAW, Direct Read After Write) that can only be recorded on once is ideal for storing large volumes of raw data, which once recorded are never changed, but are only evaluated with programs adapted to new developments. In conjunction with the appropriate software, the high storage capacity (giga byte) and the high access speed of DRAW ensure convenient data bank management. Our method of storage – one 10 minute investigation per disk – can only be recommended if the purchase costs (disk drives, software for data management) are the overriding factor, while the operating costs (purchase of disks) and personnel costs (evaluation of the individual disks) play a lesser role. The ratio operating costs/personnel costs depends on the individual project.

1.3.4. The Selection of the Pulmonary Indices Presented

The selection of the pulmonary indices presented here is based on the result of clinical-physiological considerations. Furthermore, we have only considered indices which we felt could be simplified to such a degree that they could realistically be introduced into the routine of many intensive care units and operating theaters. The question which we alone cannot answer is whether the indices that can now be easily, rapidly, and accurately measured with our system are able to influence important therapeutic decisions; in other words, whether they are *feasible, accurate, and*

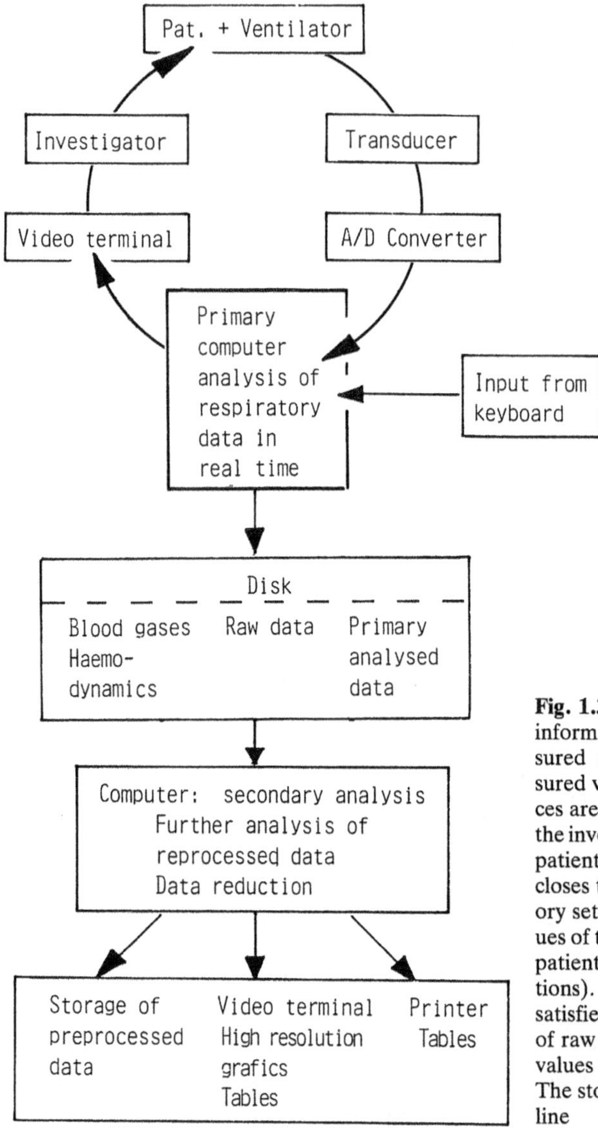

Fig. 1.3. Schematic representation of the information flow. The analog signals measured are A/D-converted. Simple measured values and pulmonary function indices are calculated on line and presented to the investigator on the video terminal at the patient's bed in real time. The investigator closes the circle by correcting the ventilatory setting according to the measured values of the last breaths and by observing the patient (e.g. watching steady state conditions). When the desired conditions are satisfied, the investigator starts the storage of raw data and digitally enters additional values (e.g. results of blood gas analysis). The stored raw data are then evaluated off line

physiologically interesting and also *clinically important.* An answer to this question will be found once such measuring systems have been installed in many units and used for many patients with a wide range of pulmonary malfunctions; i.e., after there has been sufficient *practical experience in several centers.* To achieve this goal, we are making these feasible, accurate, and physiologically interesting measuring methods available to all interested colleagues. The software is structured in such a way that the various signals (flow, pressure, concentrations) can be measured with any transducers, provided they conform to the specifications listed in Chapter 3. The software can be purchased from us for the cost of copying (about 100 Swiss francs); the only string attached is an exchange of experiences. Joint experience of several centers will show which indices will endure.

Nota Bene:

1. A clinically important index makes it possible to describe an organ function and to recognize a deviation from the normal course in order to effectively treat the patient.
2. The value of a pulmonary function index depends on the accuracy of the measurements performed on the patient and on the merits of the physiological model on which the index is based.
3. The value of an index shows up in daily routine practice, in other words, through practical experience.

2. Derivation of the Pulmonary Function Indices

2.1. Introduction

Pulmonary function indices are numbers that give a quantitative description of physiological pulmonary functions. They are derived from primary data (e.g. V_T or time courses of CO_2, pressure, etc., see Chapter 1) using certain model-concepts. The term gas exchange used here means the exchange of CO_2 for O_2 in the blood. The corresponding mass movements are brought about by pressure differences. Two types of pressure can be distinguished: absolute pressure and partial pressure (Dalton's law). While absolute pressure is the driving force of convection, partial pressure effects diffusion processes. The transport of gas between the pulmonary capillary blood and the mouth can thus be characterized by the type of driving force (Fig. 2.1) and consequently, the pulmonary function can be divided into

- breathing mechanics, purely mechanical behavior of the lung and the thorax under the influence of pressure, flow, and volume
- gas distribution within the lung (section on pulmonary volume and intrapulmonary gas mixing)
- transport of O_2 and CO_2 through the alveolar septa to the erythrocytes and blood plasma (section on transpulmonary gas transport)
- perfusion of the lung.

Although it is not possible to completely separate the two transport mechanisms (convection and diffusion), one or the other dominates, depending largely on the geometry. This aspect is discussed in detail in the section on "Lung volume and gas mixing" (see appendix 4). Supplementary information on the type and structure of the lung can be found in appendix 1.

2.2. Breathing Mechanics

2.2.1. The Patient-ventilator Unit

The mechanical behavior of the ventilator and the completely passive patient is shown in Fig. 2.2 as an equivalent electric circuit diagram. When the valve is put in position "I", the breathing cycle begins. The constant-current source supplies the $R_{tot}C_{tot}$ circuit with the inspiratory flow (V'_I). Due to the flow, the airway pressure (P_{awo})

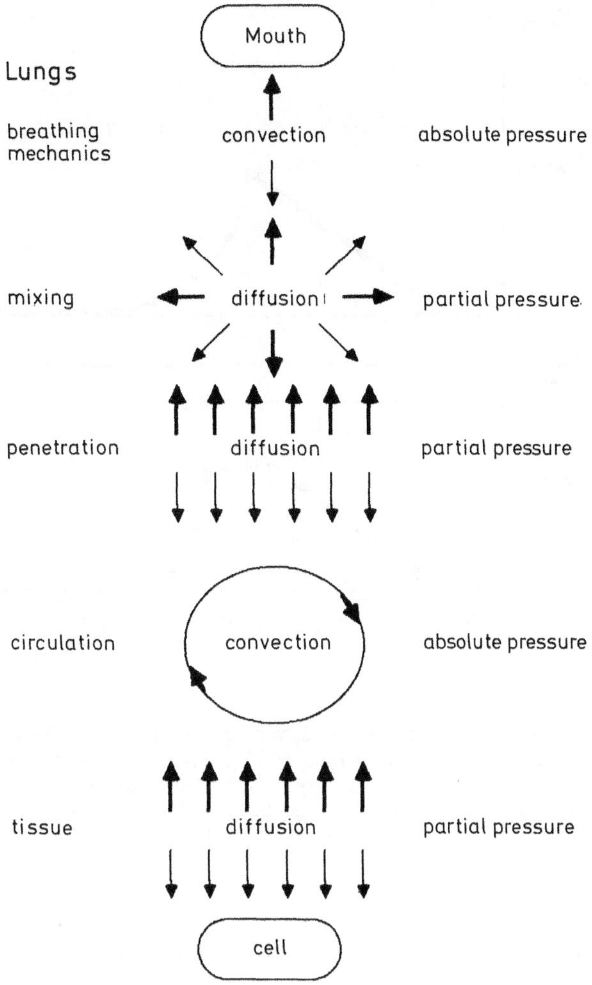

Fig. 2.1. Mechanisms that enable oxygen (thin arrows) and carbon dioxide (heavy arrows) to be transported from the mouth to the cells and back. Fresh gas reaches the lungs by means of convection, mixes with the residual gas, diffuses through various layers into the blood, and reaches the tissue by means of convection (circulation), where it diffuses into the cells. The intrapulmonary gas mixing is affected by the activity of the heart (cardiogenic mixing, Fukuchi 1977), which is not considered in this diagram

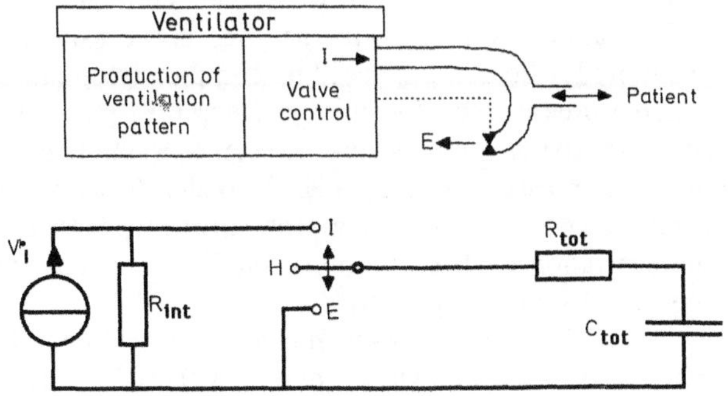

Fig. 2.2. Electric circuit diagram for the ventilator-patient unit. The ventilator consists of flow source and internal resistance R_{int}. The switch (valve with positions inspiration I, expiration E, and "hold" H) is controlled in dependence on pressure, volume, and/or time (see appendix 2: "Technical principles of mechanical ventilation")

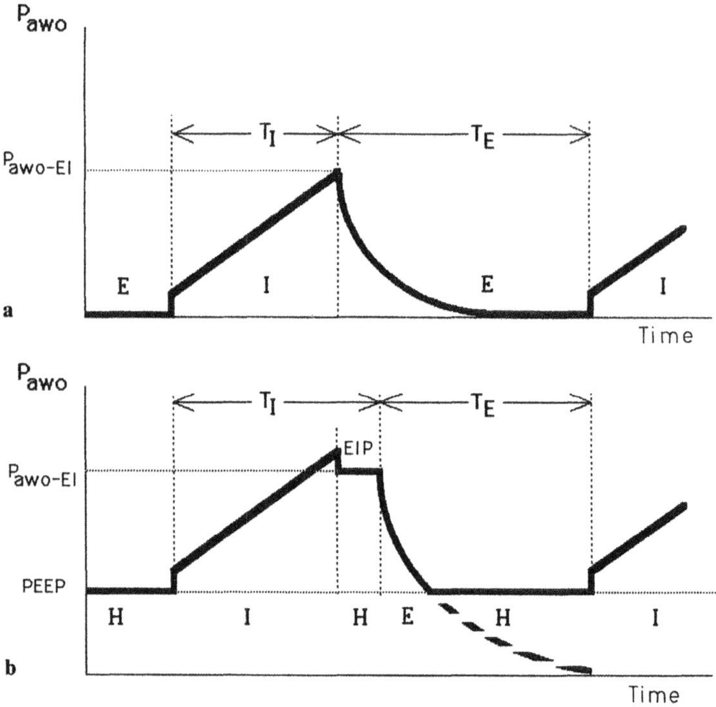

Fig. 2.3a, b. Time course of airway pressure P_{awo} in mechanically ventilated patient. T_I and T_E are inspiration and expiration times and P_{awoEI} is the end-inspiratory airway pressure. **a** Simple ventilation pattern without end-inspiratory pause EIP and without positive end-expiratory pressure (PEEP). Of the switch positions given in Fig. 2.2 only "I" and "E" are used. **b** Ventilation pattern with EIP and PEEP generated by switching to "H" at the end of inspiration (EIP) or of expiration (PEEP)

increases until the valve is moved to position "E" (Fig. 2.3a) which allows C_{tot} to be discharged. The next breathing cycle begins when the valve is put back in position "I". The positions of the valve are controlled by the ventilator in dependence on time, pressure, and/or volume (see appendix 2).

The additional switch position "H" provides differentiated patterns of ventilation (Fig. 2.3b). If "I" is not switched to "E" immediately, but "H" is selected first (fractions of up to several seconds), an "end-inspiratory pause" (EIP) results.

Expiration is only possible as long as the valve is in position "E". Putting it in position "H" before C_{tot} could be completely discharged results in a end-expiratory residual pressure, or "positive end-expiratory pressure" (PEEP). PEEP generated according to this principle has a special characteristic: it cannot be regarded as backup pressure (pressure source). Fig. 2.3b clearly shows that the respiratory pressure in expiration does not drop asymptotically towards PEEP. As soon as PEEP is reached, the expiration is abruptly interrupted. PEEP therefore cannot be interpreted as a (constant) source of pressure.

One of the main purposes of a ventilator is to provide the patient with enough gas per unit of time, known as "minute volume". This can be achieved with different ventilator settings. If a constant minute volume (MV) is assumed, the following equations result (respiratory rate RR, breath volume V_T):

$$RR = 60/(T_I + T_E) \quad [2.1]$$
$$I{:}E = T_I/T_E \quad [2.2]$$
$$MV = RR \cdot V_T \quad [2.3]$$
$$T_I + T_E = 60 \cdot V_T/V'_I \quad [2.4]$$
$$V_T = T_I \cdot V'_I \quad [2.5]$$

Changing one of the variables inevitably effects changes in others. The influence of one single quantity therefore has to be examined in connection with the variables that are dependent on it. Only V_T, V'_I, EIP, and PEEP can be set independently of one another. These are referred to in the following as *ventilation variables*. Together with the expiration time T_E, they completely define the ventilation pattern.

2.2.2. Mathematical Model

The ventilatory system in spontaneously breathing persons can be described as a set of one active and three passive sub-systems. The passive sub-systems consist of the respiratory tract with the gas volume contained in it, the lung tissue with gas volume contained in it, and the thorax. The active sub-system takes in the respiratory muscles. These muscles generate via the pleural space the driving pressure gradient, the transpulmonary pressure, which is responsible for the movement of gas in and out of the lungs.

The passive behavior of the ventilatory system can be measured by means of expanding (intrapulmonary pressure > barometric pressure) or compressing (intrapulmonary pressure < barometric pressure). If the alveolar pressure (P_A), the pleural pressure (P_{pl}), and the difference between the two ($P_A - P_{pl}$) are measured as a function of the lung volume V_L at a gas flow of zero, the static pressure-volume diagram of the entire ventilatory system – that of the thorax and that of the lungs – is obtained (Fig. 2.4). A tangent on the curves represents the volume distensibility at in

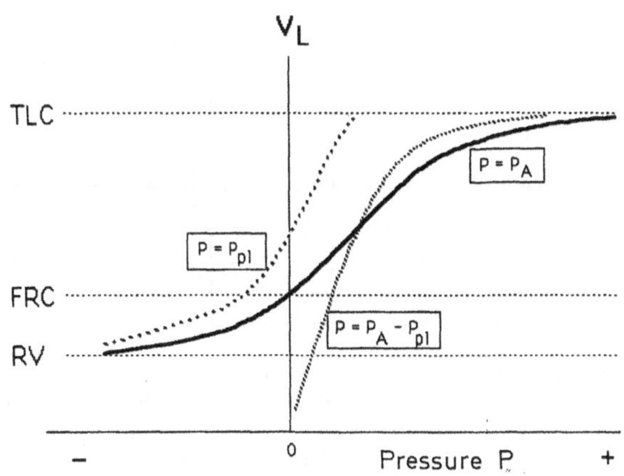

Fig. 2.4. Schematic representation of the passive behavior of thorax ($P = P_{pl}$), lungs ($P = P_A - P_{pl}$), and the entire ventilatory system ($P = P_A$). The pressure P is related to the barometric pressure. In the equilibrium position (P_A = O) the retraction force of thorax and diaphragm matches the contraction force of the lungs. The lung volume V_L is called functional residual capacity (FRC). The maximum lung volume achievable is called total lung capacity (TLC), the minimum is termed residual volume (RV). In reality, there is a small hyteresis (not shown here), which is produced by the viscoelastic tissue properties, surface tension, and surfactant.

the respective operating point and is called static "compliance". When measuring these curves, it must be taken into account that truly static conditions do not set in until after several minutes. Obviously, such measurements cannot be performed in humans. The length of the zero flow phase therefore determines the shape of the curve. A further prerequisite is that the muscles be absolutely passive – a condition which can only be fulfilled when the patient is completely relaxed or has pharmacologically induced muscular paralysis.

In mechanically ventilated subjects, the static pressure-volume curve cannot be measured during ventilation, i.e. breath-by-breath. The dynamic (short term) pressure-volume curve is measured instead. For small breaths, an equivalent electric circuit diagram can be set up under the following assumptions (Rohrer 1925, Fig. 2.5):

- pressure is homogenous and the same in all of the alveoli
- inertia and turbulences are negligible
- the elements R_{AW}, R_{ti}, R_W, C_L, and C_W are constant.

The following difference equations can thus be formulated:

$$\Delta P_{awo} = V' \cdot (R_{AW}+R_L+R_{ti}) + \Delta V \cdot (1/C_W+1/C_L) \qquad [2.6]$$

$$\Delta(P_{awo}-P_{pl}) = V' \cdot (R_{AW}+R_L) + \Delta V/C_L \qquad [2.7]$$

where Δ indicates the deviation from the equilibrium position, V' is the volume flow, V the lung volume ($\Delta V = V_T$), P_{awo} the airway pressure, and P_{pl} the pleural pressure. The equilibrium position is defined by $V' = O$ and a complete equalization of all intrapulmonary pressure gradients (P_{awo} = the alveolar pressure P_A). The total distensibility (C_{tot}) results from C_L and C_W configured in parallel ($1/C_{tot} = 1/C_W+1/C_L$).

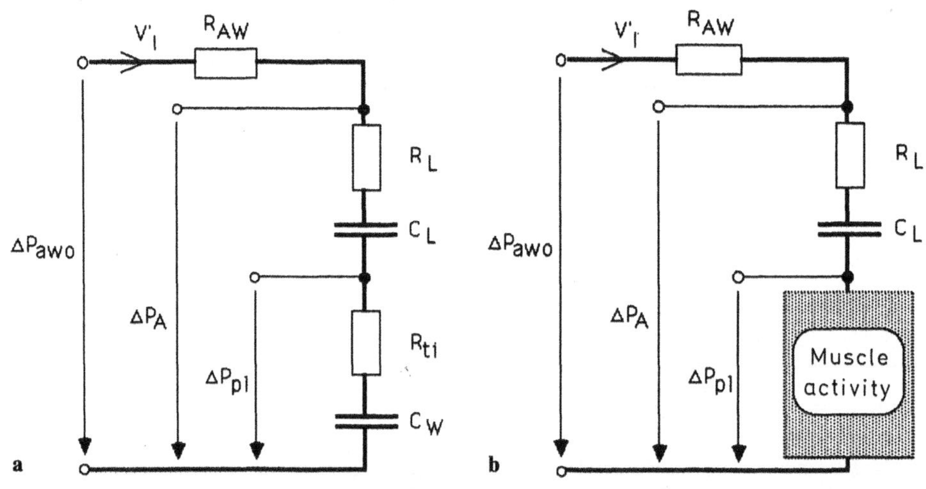

Fig. 2.5a, b. Electric circuit diagram of mechanical behavior of lungs and thorax during **a)** complete passivity of the respiratory muscles and **b)** spontaneous breathing. Δ means that these are tidal (breath by breath) processes. Only the airway pressure P_{awo} can be measured. Pleural pressure (P_{pl}) and alveolar pressure (P_A) are theoretical concepts

In interpreting the airway resistance in the ventilated adult, it must be taken into account that up to 50% of the R_{AW} (4–6 cm H_2O/l·s) is caused by the resistance of the endotracheal tube, depending on the diameter of the tube and the flow of gas in the tube (laminar, turbulent). Although the relationship between Δp and V' in the tube is not linear (Nunn 1977, p. 109), this non-linearity is considered to be negligible. The tube resistance is assumed to be *constant* and is contained in the R_{AW} values which are reported in the tables in this volume.

2.2.3. Determining Lung Compliance and Airway Resistance

To determine the model parameters from the equations 2.6 and 2.7, the pressure in the pleura has to be measured in addition to the pressure and flow in the endotracheal tube (see Chapter 3).

2.2.3a Pleural Pressure

The pressure in the pleura is inhomogenously distributed and is dependent on several factors (Agostini 1972). In other words, "pleural pressure" per se does not actually exist. Measuring the pressure distribution in the pleura is difficult and requires invasive methods, which makes it feasible only in animal experiments. Milic-Emili (1964) showed that in the spontaneously breathing human in an upright position, the esophageal pressure (P_{es}) provides a suitable substitute for a representative pleural pressure. The esophageal pressure corresponds to an average pleural pressure throughout a large portion of the pleural space. In the last few years, this measurement method has been standardized and is frequently applied. To measure the esophageal pressure, a catheter openended is inserted 45 cm deep (measured from the nostrils) into the esophagus. A deflated balloon is attached to the tip of the catheter which is filled with approximately 1 ml air after it has been positioned. The esophageal pressure can be measured in this artificially obtained space.

The method was recently tested on anesthetized spontaneously breathing patients in supine posture (Higgs 1983, Drummond 1983). According to Higgs (1983), the proportionality factor ($\Delta P_{es}/\Delta P_{pl}$) is dependent on the position of the balloon and differs from patient to patient. The error was ± 16% (SD of a group of 10 patients) in comparison with the so-called occlusion pressure (P_{awo}, which forms as a result of the effort to inhale and exhale spontaneously when the airway opening is occluded). The relationship remained linear, which makes the method suitable to measure relative changes in pressure. To arrive at a better than 16% accuracy of the absolute values, however, it seems necessary to verify the measured P_{es} with the occlusion pressure.

Since the mechanically ventilated patients were paralyzed, the occlusion test could not be performed. For this reason, a substitute-maneuver was carried out in 4 patients following heart surgery. While the tube was closed by hand, the patient's abdomen was pressed. At the same time, the P_{awo} and P_{es} were registered and subsequently compared (see appendix 3: measurement of the pleural pressure with esophageal balloon).

The results of the measurements show that P_{es} deviates from P_{pl} by an average of 30%, which confirms Drummond's values (1983). The absolute changes of P_{es} and the values derived from them for lung compliance (C_L, see below) therefore have to be interpreted with caution. Since the ratio $\Delta P_{es}/\Delta P_{pl}$ depends on the mode of breathing (see appendix), C_L tends to be underestimated during mechanical ventilation and overestimated during spontaneous breathing. The correlation between P_{es} and P_{pl}, however, is linear. Relative changes can therefore be measured precisely under the assumption that the pressure distribution in the pleura essentially remains constant during a particular breath.

In the vicinity of the esophagus, the heart produces marked pressure fluctuations which are superimposed on the pleural pressure (Fig. 2.6). To correctly estimate the model parameters, the cardiogenic oscillations have to be eliminated by filters, averaging techniques, or triggering. We use a curve fitting method which minimizes the sum of squared errors, the least squares fit method (Wald 1969, Uhl 1974).

2.2.3b Approximation of the Pressure-flow-volume Curve

Fig. 2.6 shows the projection of the three-dimensional pressure-flow-volume curves onto the pressure-volume plane. Each point corresponds to an instantaneous sample from the analog to digital converter (see Chapter 3). V and V′ of each point are put in equation 2.7 (n points per breath yield n equations). This yields a system of equations which is overdetermined, i. e. there are many solutions for C_L and $R_{AW} + R_L$. Hence an additional constraint is needed. C_L and $R_{AW}+R_L$ are determined in such a way that the sum of the squared errors becomes minimal. The solid lines in Fig. 2.6 show the synthesized pressure-volume curves which were generated using equation 2.7, the

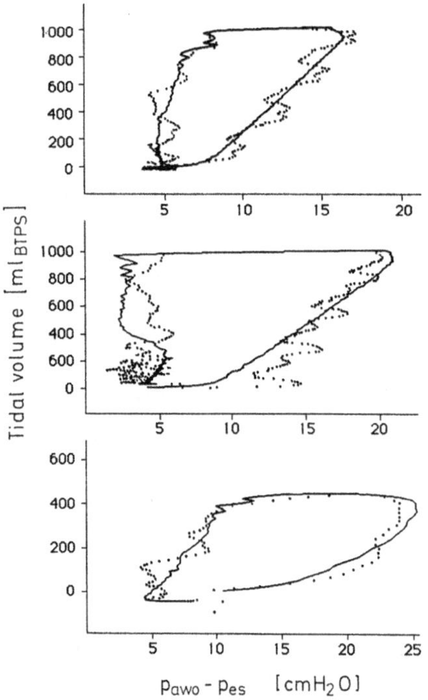

Fig. 2.6. Measured pressure – volume curves (dotted lines) and superimposed approximations according to the least squares fit method (solid lines). The curves run counterclockwise with inspiration in each chart beginning at the bottom left. The upper two curves were measured during mechanical ventilation, the bottom one during spontaneous breathing. The approximation is adequate, though systematic deviations can be seen in the middle curve

estimated C_L and $R_{AW} + R_L$ as well as flow and volumes to back-calculate the pressures.

The parameters C_{tot} and $R_{AW}+R_L+R_{ti}$ are determined in the same manner but using equation 2.6.

Although it is known that in the normal case R_{AW} is much greater than R_L, the sum of $R_{AW} + R_L$ cannot be divided any further. This is because the alveolar pressure P_A is a theoretical concept and not a measurable quantity. In patients who are not totally paralyzed or who are breathing spontaneously, the model in Fig. 2.5a must be replaced by the one in Fig. 2.5b, since the muscles are no longer passive but are at least partially active. The chest wall no longer acts as a passive component. Consequently, C_{tot} is not defined and therefore cannot be determined. The lung compliance C_L nevertheless can be determined, since it is not affected by chest wall movement.

2.2.4. Influence of the Mechanics on the Distribution of the Ventilation

The difference between alveolar pressure (P_A) and pleural pressure (P_{pl}) is the transpulmonary pressure. This pressure is not homogenous throughout the lungs, but increases starting at the "bottom" (dependent regions) and moving "upwards" (non-dependent regions). One of the reasons for this inhomogeneity is the weight of the lung tissue. If for each alveolus a pressure-volume characteristic is assumed, as is represented in Fig. 2.4, the pressure gradient effects (when the person is standing upright) a different resting distensibility of apical and basal alveoli (different operating points in the pressure-volume diagram). This results in a sequential and inhomogenous ventilation, since the basal pulmonary regions, thanks to the favorable operating point, receive a considerable amount of fresh gas while the apical areas, which are already inflated, receive only little (Milic-Emili 1966).

More recent investigations (Agostini 1972) have shown that the deformation of the lungs is another cause of non-uniform ventilation. The lungs and the chest have different geometric shapes. Since the lungs are easily deformable, they are sort of squeezed into the chest cavity. The resulting deformations cause changes in surface tensions and these in turn affect the distribution of the ventilation. Thus, the lungs have not one but many operating points in the pressure-volume diagram and also various distensibilities which are distributed around a mean value. For this reason, the values C_L, R_{AW}, and C_{tot} determined above do not represent *single* discrete components, but are weighted mean values of distributions.

Nota Bene:

- Both convection and diffusion take part in the transport of gas. In the different segments of the airways and the lung (and of the circulation and the metabolizing peripheral cells), the geometric arrangement largely decides whether convection or diffusion prevails as transport mechanism.
- The patient and the respirator form a functional unit.
- C_{tot}, C_L, and R_{AW} can be determined breath-by-breath and without any special maneuver (e.g., without an end-inspiratory pause) and without interruption of the respiration or ventilation.

- C_{tot}, C_L, and R_{AW} are not discrete components, but are (weighted) mean values of distributions.
- C_{tot}, R_{tot}, P_{awo}mean, and P_{awo}max can only be defined when the breathing muscles are in an absolutely passive state (muscular paralysis). Values for C_{tot} and/or R_{tot} determined while the muscles are active are not valid.
- The pleural pressure is a physiological concept and does not exist as a real quantity. The esophageal pressure as a substitute for the pleural pressure for intensive care patients is of only limited value. The difficulties encountered are not caused by the superimposed cardiogenic artefacts (these can be eliminated by using proper techniques), but by the inhomogenous and variable distribution of the pleural pressure.

2.3. Lung Volume and Intrapulmonary Gas Mixing

2.3.1. Nitrogen Washout

To measure the lung volume, the patient is given a gas without nitrogen at the start of an inspiration. In other words, no nitrogen is breathed in, however, the intrapulmonary nitrogen is gradually breathed out, or washed out, of the lungs (see Fig. 1.2, Chapter 1). The lung volume can be determined by measuring the total quantity of eliminated nitrogen. The course of the nitrogen washout provides information on the intrapulmonary gas mixing.

Technical difficulties arise during the application of the nitrogen washout in ventilated patients due to the different gas viscosities, which cause the volume delivered by the ventilator to change when the gas mixture is changed. To get around these difficulties, a ventilator was modified. The Monaghan M-250 proved to be well suited for this purpose, as it has two separate bellows which are simultaneously operated from one single pressure generator (Fig. 2.7). These bellows were separated so they could be supplied by two different sources of gas. A switching valve to change from one bellows to the other was also built into the ventilator (Nyffeler 1984). When the switch is activated, the gas interface reaches the mouth of the patient after an inspiration of V_{ext}.

The best gases for washout procedures are N_2-free gas mixtures, which ideally have no interaction with the blood and the lung tissue (so-called "inert" gases). Pure oxygen is not suitable, because it reacts with blood, can lead to absorption atelectases in spontaneously breathing patients, and influences – at least potentially – the hemodynamics (e.g. hypoxic vasoconstriction). For our studies we primarily used argon/oxygen mixtures. The few washout maneuvers with pure oxygen are marked and were used only to measure the lung volumes. In such cases, the washout curves were not analyzed.

2.3.2. Measurement of the Functional Residual Capacity (FRC)

FRC is defined as the intrapulmonary gas volume at the end of a normal expiration (see Fig. 2.4) and can vary breath by breath in the spontaneously breathing patient.

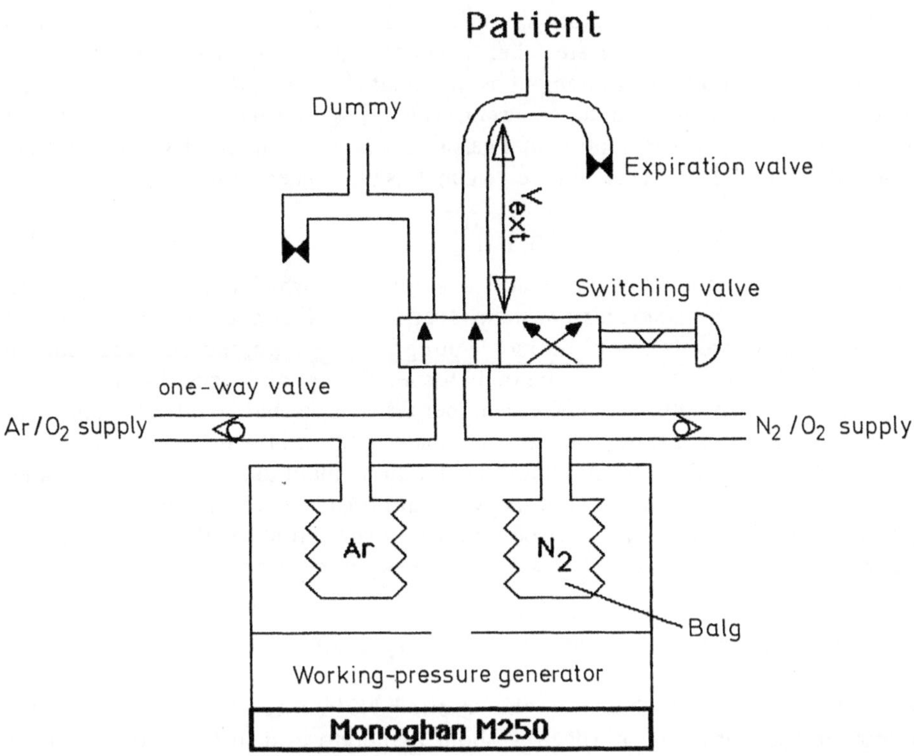

Fig. 2.7. Diagram of the modified Monaghan M 250 ventilator for the purpose of carrying out the N_2 washout in ventilated patient. Since both bellows are operated from the same pressure generator they produce the same ventilation pattern. Activating the switching valve starts the N_2 washout. Nitrogen-free gas reaches the mouth of the patient after volume containing N_2 (V_{ext}) has been inspired. This volume must be taken into consideration in all of the calculations

The FRC can be measured by the washout method, insofar as the entire intrapulmonary gas volume is accessible to the gas mixture. This condition, however, is not necessarily met (trapped air caused by mucous plugs, for instance) and the lung volumes measured by the nitrogen washout method therefore cannot be considered FRC. For this reason, a new term is defined:

Definition: the nitrogen washed out in a certain period is proportional to the pulmonary gas volume accessible in this period. This gas volume is therefore termed "accessible pulmonary (gas) volume" (APV).

The accumulated, total expired nitrogen volume is a measure for the APV. If the nitrogen concentration in the lungs is homogenous at the beginning of the washout ($F_{N_2}{}^{\circ}$), the following equation applies for the total washed out nitrogen volume V_{N_2tot} (Eq. 2.8):

$$V_{N_2tot} = F_{N_2}{}^{\circ} \cdot APV - V_{N_2rest} \qquad [2.8]$$

V_{N_2rest} is the volume of nitrogen which remains in the lungs due to incomplete washout.

Both inspiratory and expiratory measurements are taken, so that any re-breathed nitrogen volumes (e.g. V_{ext}) are taken into account in V_{N_2tot} (see Chapter 3 and appendix). When the measurement is terminated, the end-expiratory nitrogen concentration has dropped to a certain value F'_{N_2}. Using F'_{N_2}, V_{N_2rest} can be determined under the assumption that gas mixing is complete in APV, i.e., that the total APV after the last measured breath contains the concentration F'_{N_2}. The APV is then

$$APV = V_{N_2tot} / (F_{N_2}{}^{\circ} - F'_{N_2}) \qquad [2.9]$$

The assumption made above introduces a minor error, which becomes smaller with decreasing F'_{N_2}, i.e., the longer a washout is carried out. However, it is not possible to continue the N_2-washout until no more nitrogen can be washed out. The limiting factor is the nitrogen content in the body tissue, since the nitrogen dissolved in the body is washed out only very slowly. Another difficulty is that no differentiation can be made between very slow pulmonary compartments and the nitrogen in the body. To calculate the APV, the simplified assumptions mentioned above were made. Because the maximum duration of the washout procedure was limited to 7 minutes (capacity of a floppy disk), we dispensed with a correction for the nitrogen in the body.

2.3.3. Gas Mixing in the Acinus: The Stationary Interface

For nitrogen to be washed out, the inspired N_2-free gas has to mix with the intrapulmonary gas. The nitrogen washout is therefore ideal for studying mixing processes. In the following we will discuss how this mixing happens to take place.

First of all, only one single acinus is taken into consideration. For purposes of simplification, it is assumed that this acinus is rigid and that it possesses the dimensions given in the appendix Fig. A1.4 (thumbtack model). An ideal source of nitrogen is assumed to be located at the distal end of this acinus. Let us consider first only convection: at the entry of the acinus a constant inspiration flow V′ is produced. The velocity of convection (v) of the gas molecules at the entry of the acinus is then $v = V'/S$ (S represents the cross-sectional area the flow passes through). According to the law of continuity, the flow has to remain constant in the acinus. Since the cumulative cross-sectional area increases considerably towards the end of the acinus, the convection velocity decreases (Fig. 2.8). Diffusion, on the other hand, is increased by large concentration gradients and large exchange areas. It decreases when the distances to be covered become longer. The diffusion velocity thus decreases towards the mouth (Fig. 2.8).

The direction of the diffusive mass transport is opposite to that of the convective mass transport. A stationary interface (SI) is established at the point where diffusive mass transport compensates for convective mass transport. This point is mathematically defined as the site of the largest concentration gradient in a steady state ($\delta F_{N_2}/\delta t = O$ and $\delta^2 F_{N_2}/\delta r^2 = O$; (see appendix 4, “transport – equation for convection and diffusion”). The SI represents the transient between convection and diffusion and thus the proximal limit of the alveolar space. It is dependent on the inspiratory flow, the diffusion constants, and the geometry. It is independent of the concentration gradients (Paiva 1972, see appendix 4) and of the volume which flows into the acinus.

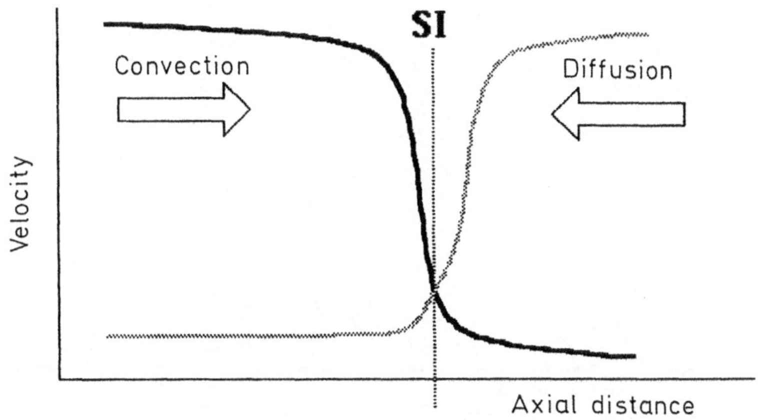

Fig. 2.8. Qualitative representation of the convection velocity (solid line) and diffusion velocity within one acinus during constant inspiratory flow. N_2-free gas flows towards the end of the acinus. N_2 diffuses from the acinus. The stationary interface (SI) is located where the diffusive mass transport compensates for convective mass transport

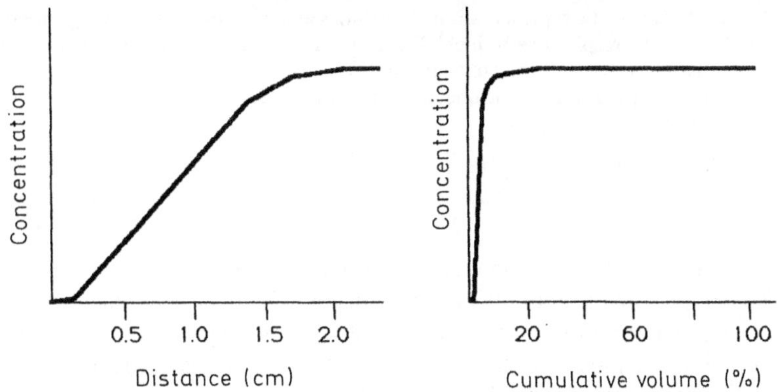

Fig. 2.9. Concentration profiles in the thumbtack model of the acinus 1 second after the beginning of inspiration with N_2-free gas (from LaForce 1970). The course of N_2 in the left diagram is plotted versus the axial distance from the entrance up to the acinus (terminal bronchioles, see appendix 1: Morphology) and in the right diagram vs. the cumulative volume in this acinus

Model calculations yielded concentration profiles in the acinus as shown in Fig. 2.9 (from LaForce 1970). Apparently, how the F_{N_2} is plotted, namely vs. axial distance or vs. volume, determines whether the concentration course is interpreted as "interface" or as a "blurry" concentration profile. Since volumes are measured at the mouth, it seems justified to assume the concentration profile in the acinus as interface.

2.3.4. Gas Mixing in the Lungs

The observations made on one single acinus can be applied to the entire lungs, which can be regarded as a tree-shaped, branching structure of gas-filled spaces (Fig. 2.10).

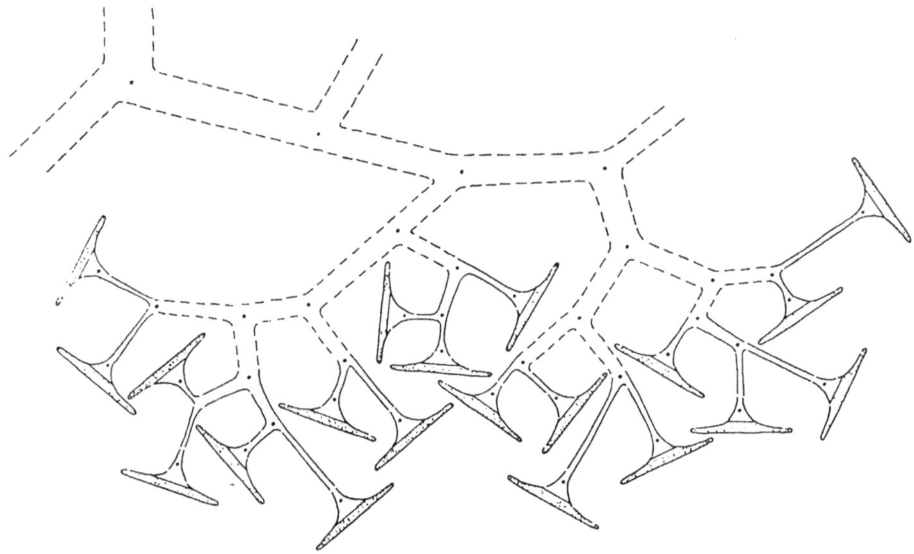

Fig. 2.10. Schematic representation of the lungs as a tree-shaped branching structure with bronchioli (dashed lines), respiratory bronchioli (solid lines), and acini ("trumpets") as terminal parts. This model assumes that at the end of an inspiration a stationary state and thus a stationary interface forms in each acinus (indicated by the lines in the acinus). The gas distal to the interface (doted area) is called "alveolar gas"

The entrance to the trachea is the origin and the acini are the terminal parts of this structure. Several generations of branches arise from the trachea and lead to the acini. That means that there are pathways of various lengths leading from the trachea to the individual acinus. The length of all of the paths is statistically distributed around a mean value (Weibel 1963).

During inspiration, an SI forms in correspondence with the flow in this acinus. The volume contained in all of the airway spaces from the mouth to the SI's is called airway-dead space (V_{dAW}) or series dead space (V_{ds}). This "space" is not anatomically fixed but is functionally variable. Moreover, analogous to the distribution of the path lengths, a distribution of series dead spaces must be assumed.

As the gas composition within the lung cannot be measured, the expirate is interpreted as being representative of the intrapulmonary conditions at the end of the previous inspiration. During expiration, gas mixing takes place at each branch of the bronchial tree. When the SI's are washed out of the acinus, the concentration of nitrogen is increased each time two SI's merge into the same branch. The SI with the highest flow and concentration exerts the most influence. The course of nitrogen at the mouth (Fig. 2.11) therefore has to be interpreted as flow-, volume-, and concentration-weighted cumulative distribution of all SI's (Horsfield and Cumming 1968) and is plotted against the expired volume. The nitrogen concentration exhibits (according to Aitken 1928) three phases: pure deadspace (phase I), increased alveolar concentration (phase II), and alveolar plateau (phase III). Under certain circum-

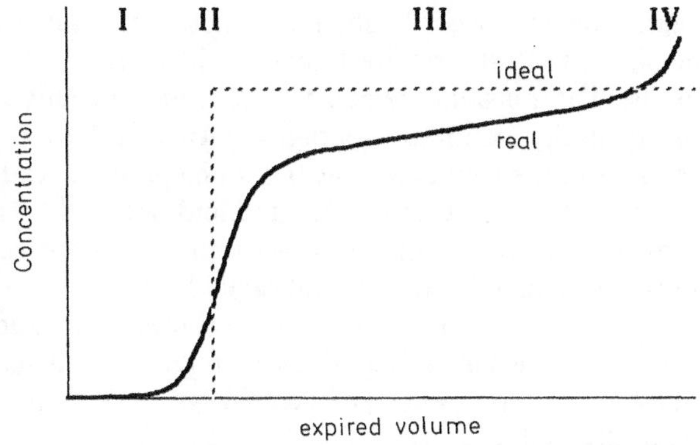

Fig. 2.11. Expiratory course of N_2 in washout (ordinate) plotted against the expired volume (V_E, abscissa). The ideal curve is represented as dashed lines. In Phase I only gas from the airway dead spaces is expired. In the ideal case, phase I leads directly into Phase III, the alveolar plateau, which is horizontal. In reality, the curve is severely deformed and occasionally there is an additional Phase IV

stances (special breathing maneuvers), a fourth phase can be observed, which is called the closing volume.

Closing volume

The single breath diagram for nitrogen sometimes shows a distinct up-slope of phase III which terminates the alveolar plateau. Following the convention which defines phases I, II, and III, this is called phase IV. To study the phenomenon, a maneuver was introduced consisting of 1) exhaling down to residual volume, 2) then taking a deep breath with nitrogen-free gas up to total lung capacity, and 3) exhaling slowly back down to residual volume. The nitrogen content is then plotted against expired volume. The expired volume axis is converted to a lung-volume axis by substraction from the end-inspiratory volume (total lung capacity). Finally, the volume at which phase IV begins is estimated by hand or by computer (Craven, 1976).

Although the exact mechanism for a phase IV is unclear (Cumming 1984), the following hypothesis leads to the term "closing volume": at residual volume, the alveoli at the top of the lungs are still well inflated, while in the dependent parts of the lungs they are already smaller. Consequently, during the following inspiration the dependent parts of the lungs receive more of the ventilation and more of the nitrogen-free gas. As a result, the nitrogen in the dependent regions is diluted more than the nitrogen in the independent regions. This causes a vertical concentration gradient. During the subsequent expiration, the contribution of the dependent regions to expiration flow ceases first which produces phase IV. Functionally, these regions are now "closed" since gas flow has stopped.

The volume of the lungs at which the transition from phase III to phase IV indicates airway closure is called "closing volume" (Milic-Emili 1966, McCarthy 1972, Engel 1975).

Closing volume in the strict sense can only be studied in a spontaneously breathing, highly cooperative patient. One could, however, try to imitate this investigation in the unconscious and even paralyzed patient. The active voluntary elements would have to be substituted with passive ones imposed on the patient by the ventilator. Obviously, exhaling down to residual volume would have to be imitated by negative pressure expiration and, accordingly, inspiring to total lung capacity by high positive pressure inspiration. Both volume limits would therefore have to be replaced by pressure limits. The question remains as to which negative and positive pressures of the mechanically ventilated patients correspond to residual volume and total lung capacity of the spontaneously breathing patient. And perhaps more importantly: can lungs which are artificially inflated and deflated by means of positive and negative airway pressure be compared with lungs that are gently expanded and reduced by pressure changes in the pleural cavity?

The closing volume used to be measured often and many interesting observations were made such as its dependence on gravity and age. However, the closing volume does not appear in recent clinical literature. It seems the index did not turn out to be clinically useful.

In light of these questions we abstained from trying to estimate the closing volume in ventilated patients.

2.3.5. Measurement of the Airway Dead Space (Series Dead Space)

Phase II is a cumulative distribution of SI appearing at the mouth. The distribution can be described by means of a mean value, variance, and shape. That means that the larger the volume, the sooner its contribution is transferred from phase II to phase III, where it can no longer be differentiated from parallel (i. e. alveolar) dead spaces. It is therefore impossible to measure the total airway dead space. If the change in concentration (dF/dV) is plotted against the expired volume, (V_E) (Fig. 2.12), it shows the flow-, volume-, concentration-, and sequence-weighted distribution of all the alveolar gas interfaces measurable at the mouth. The mean volume of this distribution is defined as the measurable portion of the series dead space or as Pre-Interface Expirate (PIE) (Wolff 1984, Brunner 1985a).

PIE is calculated in two steps:

1. The analysis is focused on phase II. The volume in which the N_2 concentration exceeds 50% of the end-tidal concentration of the same expirate (V_a in Fig. 2.12) is doubled. Only this interval is considered for the further calculations.
2. The first derivation $\Delta F_i/\Delta V_i$ is determined for each instantaneous sample i in the interval under consideration (eq. 2.10). PIE is calculated as the mean value of the distribution of all dF/dV_i (first moment, eq. 2.11, summation over all samples i).

$$\Delta F/\Delta V_{i+1} = F_{i+1} - F_i / V_{i+1} - V_i \qquad [2.10]$$
$$PIE = 1/\Sigma\Delta F_i \cdot \Sigma V_i + \Delta V_i/2 \cdot \Delta F_i \qquad [2.11]$$
$$V_A = V_T - PIE \qquad [2.12]$$

PIE is also an estimated value for V_{dAW} (also known as V_{dS}) and allows the measurement of the alveolar ventilation V_A (eq. 2.12).

In lung physiology, V_A is usually calculated per minute and has the dimension of a flow (V'_A). Dividing by APV yields the relative, alveolar ventilation of the pulmonary volume. The relative ventilation of V_{dAW} is infinite, as its residual volume in each single breath is totally washed out or exchanged. V_{dAW} is thus a space that "diverts" the inspirate which would otherwise be available for gas mixing in the alveolar space.

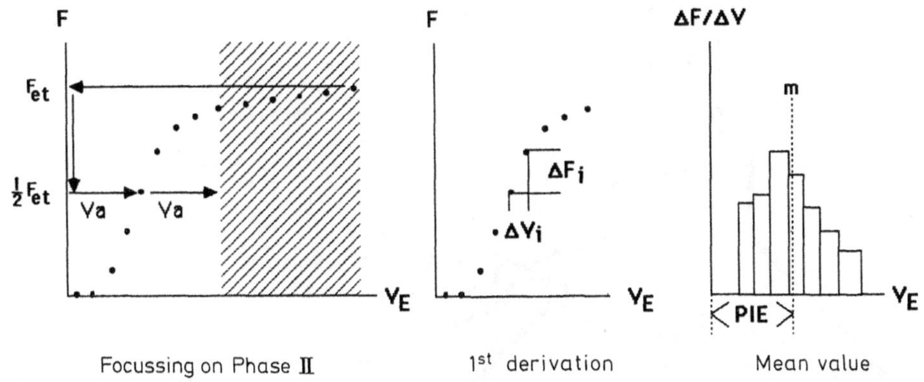

Fig. 2.12. Determination of PIE. Focusing on phases I and II (N_2 concentration F against expired volume V_E) with the aid of the endtidal concentration (F_{et}). The shaded part is not included in the further analysis. The first derivation of all $\Delta F_i/\Delta V_i$ yields a flow- and concentration-weighted distribution function. Its mean value m is defined as pre-interface expirate (PIE)

It is a mixing dead space and therefore also cannot participate in the exchange of gas with blood.

2.3.6. Washout Efficiency from the Washout Curve (Decay Curve)

Each lung has mixing dead spaces, even in the alveolar space (Fig. 2.13). Inhomogenous ventilation has the functional effect of alveolar dead space. In terms of function, this means that only a part of the ventilation (V′) is available for mixing within the lung volume. The alveolar dead space ventilation (V'_{dA}) is characterized by an infinitely large relative ventilation. The ideal lung has no dead space, inspired gas is equally distributed to all alveoli. The ideal lung can therefore be replaced by a single compartment with perfect mixing.

Definition: gas mixing in the alveolar space is considered perfect when all the alveolar gas is completely mixed with the inhaled fresh gas (equilibrium).

In the following, the efficiency of the N_2 washout is derived by comparing the measured values obtained in the patient with the values expected from the ideal lung (100% washout efficiency). It is assumed that V_T, APV, and PIE do not change during the nitrogen washout. The N_2 concentration in the ideal lung is calculated using mass balance equations: equation 2.13 for the first (partially) N_2-free breath, equation 2.14 for the nth breath:

$$F_{ideal}(1) = F_{N_2}^{O} \cdot [APV + V_{ext}]/[APV + V_T - PIE] \quad [2.13]$$

$$F_{ideal}(n) = F_{ideal}(n-1) \cdot APV/[APV + V_T - PIE] \quad n = 2, 3, \ldots \quad [2.14]$$

where $F_{N_2}^{O}$ is the initial concentration, V_{ext} is defined in Fig. 2.7. From this, the amount of nitrogen remaining in the lung after each breath can be computed (eqs. 2.15 and 2.16):

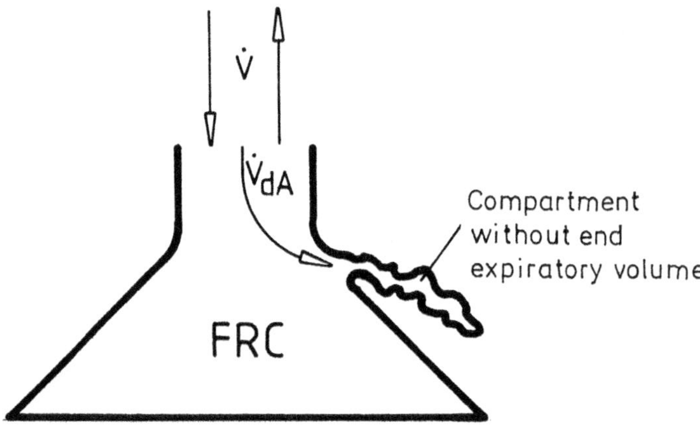

Fig. 2.13. Parallel model of the mixing dead space

$$V_{N_2rest}(1) = (APV + V_{ext}) \cdot F_{N_2}{}^{\circ} - (V_T - PIE) \cdot F_{ideal}(1) = APV \cdot F_{ideal}(1) \quad [2.15]$$

$$V_{N_2rest}(n) = V_{N_2rest}(n-1) - (V_T - PIE) \cdot F_{ideal}(n) = APV \cdot F_{ideal}(n) \quad [2.16]$$

V_{N_2rest} is divided by the total amount of intrapulmonary nitrogen ($APV \cdot F_{N_2}{}^{\circ}$) and is semilogarithmically plotted breath-by-breath in % against the so-called turn over number (TO) (Fig. 2.14), which is defined by the equations 2.17 and 2.18:

$$TO(1) = (V_T - PIE - V_{ext}) / (APV + (V_{ext}) \quad [2.17]$$
$$TO(n) = TO(n-1) + (V_T - PIE)/APV \quad n = 2, 3 \ldots \quad [2.18]$$

The resulting straight line is called an *ideal decay curve*. Its washout efficiency is 100% and hence it is steeper than the measured decay curve (Fig. 2.14). The washout efficiency (WO.eff. in %) is computed by comparing the number of turnovers after 90% of all the nitrogen is washed out (eq. 2.19, as suggested by Prowse 1973):

$$WO.eff. = TO_{ideal90\%} / TO_{real90\%} \cdot 100 \quad [2.19]$$

Since the breaths rarely lie on the 10% line, the theoretical points of intersection are calculated as follows: for both the ideal curve and the experimental curve, 4 of the points closest to the 10% limit are each approximated by a single exponential curve (least squares fit). The points of intersection of the curves with the 10% line (TO_{ideal} and TO_{real}) are calculated from this.

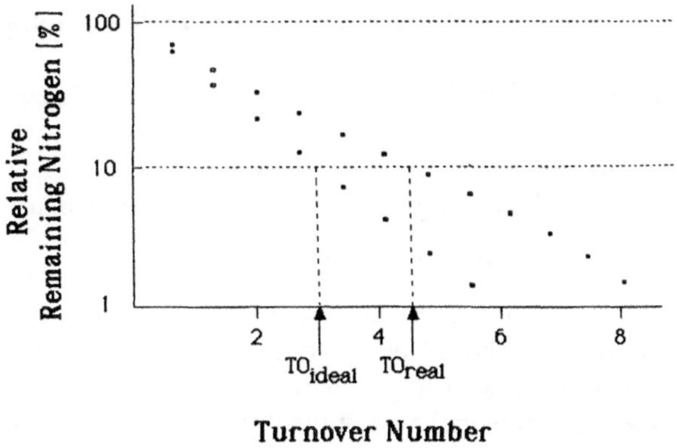

Fig. 2.14. Nitrogen decay curve of a ventilated patient with $V_T = 1000$ ml and APV = 1500 ml. The amount of N_2 expired during an N_2 washout is subtracted breath-by-breath from the total amount of N_2, is divided by the total amount of N_2 (relative remaining nitrogen), and semilogarithmically plotted against the "turnover number" (TO). The ideal curve is steeper than the real curve and reaches the 10% limit after TO_{ideal}. In reality, additional breaths are required before 90% is washed out (TO_{real})

Nota Bene:

- Gas mixing and gas distribution in the lung can be studied non-invasively using the N_2 washout procedure.
- The lung volume can also be measured, however, only that volume accessible for gas mixing during the period of the washout (APV). That is why APV can differ from the lung volume determined in the whole body plethysmograph, which measures total intrathoracic gas volume (ITGV).
- During inspiration an interface forms within the acinus which in the simplified model does not move in steady state (stationary interface = SI).
- Although there is little doubt that there is a "blurry" concentration gradient in the alveoli with respect to axial distance, an abrupt concentration interface can be assumed with respect to volume.
- The position of the stationary interfaces depends on anatomy and function.
- The N_2 single breath diagram contains information on the different relative ventilations in the lungs, on intrapulmonary gas mixing, and on the expiratory flow pattern.
- Phase II of the N_2 diagram can be interpreted as a cumulative distribution function. PIE is the weighted mean value of its density and an estimation of the airway dead space.
- Tidal volume minus series dead space volume reaches the alveolar zone: alveolar ventilation.
- In the decay curve, the portion of nitrogen that is not yet washed out is plotted against the cumulated alveolar ventilation. From this, an index of the quality of the gas mixing and gas distribution in the alveolar space, the so-called washout efficiency, can be calculated (WO.eff.). It should be noted that the crucial step in this procedure is the definition of the ideal decay curve.

2.4. Transpulmonary Gas Transport: Exchange of CO_2 and O_2

2.4.1. Diffusion from and into the Pulmonary Capillary Blood

Blood and gas are separated by septa and interstice. This barrier is overcome by diffusion. Fick described the transport process in a steady state as follows (eq. 2.20):

$$dM/dt = -D \cdot F \cdot \delta C/\delta x \qquad [2.20]$$

where dM/dt = mass flow, C = concentration, F = exchange area, D = diffusion coefficient. The diffusion coefficient D is primarily dependent on the type of gas (Graham's law $D_1/D_2 = \sqrt{(m_2/m_1)}$, m = molecular weight of gas 1 and 2, respectively), on the thickness of the diffusion barrier, and on the temperature. The driving concentration gradient $\delta C/\delta x$ is proportional to the partial pressure difference between the alveolar gas (P_A) and the mixed venous blood ($P_{\bar{v}}$). The proportionality factor (between partial pressure and content) is the solubility coefficient of the respective gas. Carbon dioxide and oxygen diffuse from the gas phase into the blood phase and vice-versa until the partial pressure is equal on both sides of the diffusion barrier. For the following considerations, we will assume that this equalization is instantaneous, in other words, that there are no diffusion limitations.

2.4.2. Dissociation of O_2 and CO_2 in the Blood

The O_2 and CO_2 content in the respiratory gas is directly proportional to the corresponding partial pressures. The proportionality factor depends only on the barometric pressure and on the water vapor pressure. This does not apply for the gas content in the blood, since gas is not only physically dissolved but is also chemically reversibly bound. The relationships between content and partial pressure are non-linear and mutually dependent.

O_2-dissociation curve: oxygen is carried by the blood in two different forms: physically dissolved and chemically bound to hemoglobin. The relationship between partial pressure and content is non-linear and, in the presence of high partial pressures, very flat (Fig. 2.15). The O_2-dissociation curve is shifted as a result of variations in temperature, CO_2 partial pressure (Bohr effect), and pH. Kelman (1966) wrote a computer algorithm for the approximation of the O_2-dissociation curve, which is used here for calculating the O_2 content (CO_2 in vol%) from the O_2 partial pressure (see list of formulas, appendix 8).

CO_2-dissociation curve: carbon dioxide is carried by the blood in a physically dissolved and a dissociated form. In contrast to oxygen, the CO_2 dissociation curve is very steep in the physiologically relevant range. That means that the slightest change in partial pressure effects a significant alteration of the CO_2 content (Fig. 2.16). CO_2 content, pH, and buffer action are in close relationship. The reversible chemical dissociation is dependent on temperature, pH, hemoglobin, and O_2-saturation (Haldane effect). There is also an algorithm by Kelman (1967) for the CO_2 dissociation curve, which is used here in its original form for calculating the CO_2 content (CO_2 in vol%) from the CO_2 partial pressure (see list of formulas, appendix 8).

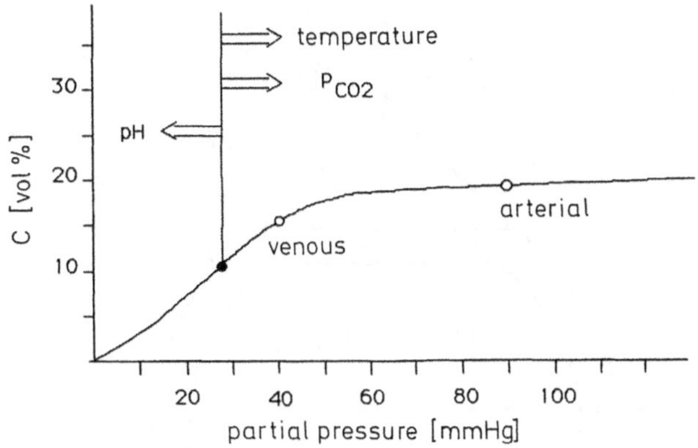

Fig. 2.15. O_2 dissociation curve. The O_2 content in the blood (C_{O_2}) is plotted against the O_2 partial pressure (P_{O_2}) with a hemoglobin content (Hb) of 13 mg/100 ml. Part of the O_2 is physically dissolved, which is why the curve is not flat when the P_{O_2} is high but continues to rise slightly. Temperature, pH, and P_{O_2} (Bohr effect) shift the curve in the direction shown by the arrows. The points "venous" and "arterial" mark the O_2 content in the normal subject (breathing ambient air)

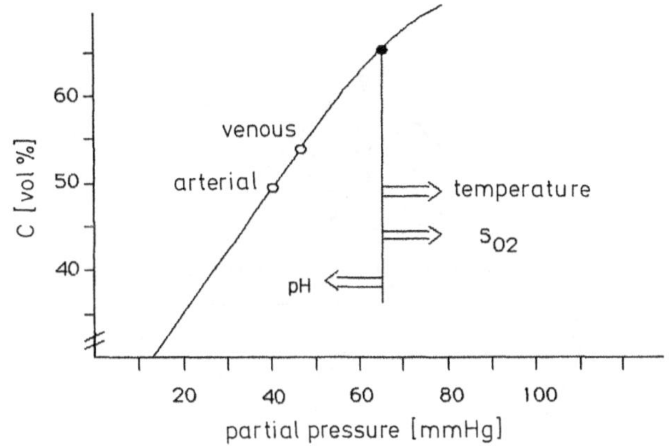

Fig. 2.16. Schematic diagram of CO_2 dissociation curve. The CO_2 content is plotted against the CO_2 partial pressure (P_{CO_2}). In the physiological range the curve is much steeper than that for O_2. The arrows indicate in which direction the curve shifts when there is an increase in temperature, pH, and O_2 saturation (S_{O_2}, so-called Haldane effect). The points "venous" and "arterial" mark the CO_2 content in the normal subject (breathing ambient air)

2.4.3. The "Riley" 3-compartment Model

The ideal lung without airway dead space is schematically represented in Fig. 2.17a. This lung is homogenously ventilated with V' and perfused with Q'_c, i.e., there is only one ventilation-perfusion relationship, which is expressed as ventilation/perfusion ratio (V'_A/Q').

In the ideal case shown in Fig. 2.17a, the effective alveolar ventilation V'_{eA} corresponds to the total ventilation (V') and the capillary perfusion Q'_c to the total perfusion (Q'). It is thus easy to measure the alveolar partial pressures of O_2 (P_{AO_2}) and CO_2 (P_{ACO_2}). They are identical with the arterial partial pressures P_{aO_2} and P_{aCO_2} and the expiratory partial pressures P_{EO_2} and P_{ECO_2}, respectively.

In the ventilated patient, ventilation and perfusion do not ideally match up. Upper regions (non-dependent) are ventilated more (V'_A/Q'high) and the bottom (dependent) areas are perfused more (V'_A/Q'low). These V'_A/Q' inhomogeneities result in reduced efficiency of the gas exchange, i.e., only fractions of ventilation and

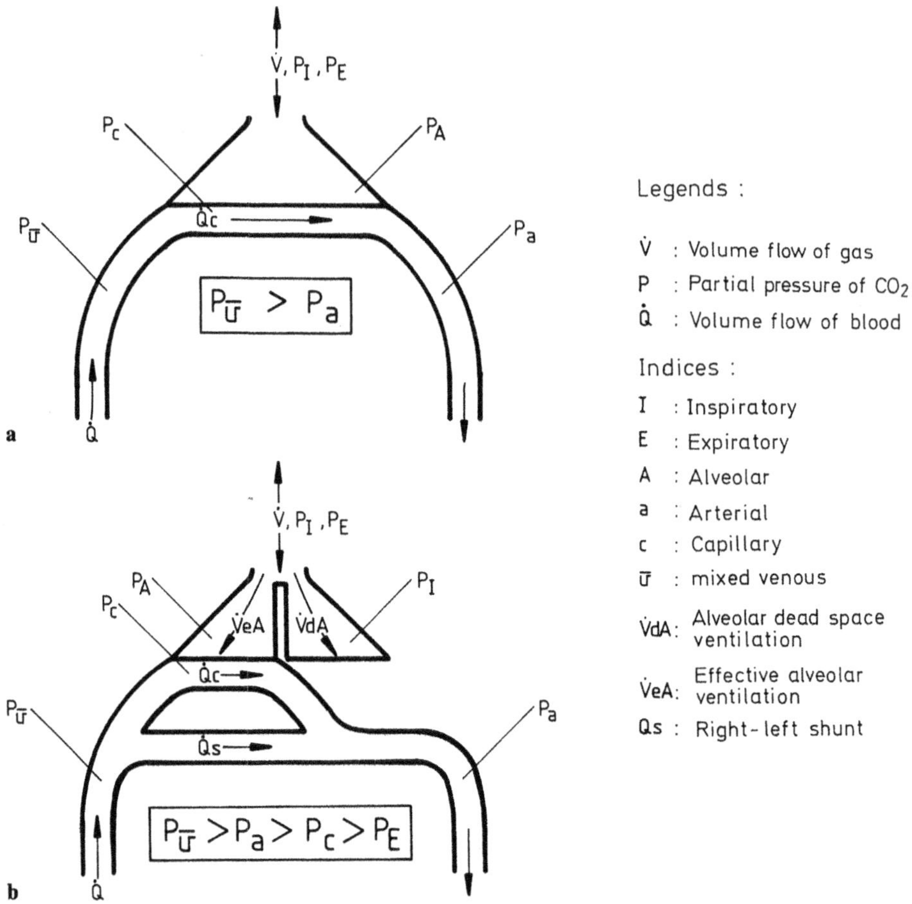

Fig. 2.17a, b. Lungs and bronchial circulation as a V'_A/Q' model with P_{CO_2} gradients given. An ideal mixture is assumed within the compartments, the partial pressure equalization over the blood-gas barrier is instantaneous. **a** Ideal case, homogenous with lung ventilated with V' and perfused with Q'. V' is identical to the effective alveolar ventilation V'_{eA} and Q' is identical to the capillary perfusion Q'_c. **b** Classical three-compartment model. Dead space ventilation (V'_{dA}) and shunt (Q'_S) do not contribute to the gas exchange

perfusion remain "ideally" effective. This inefficiency can be functionally described by two additional compartments (model by Riley and Cournand 1949):

1. one dead space that is ventilated but not perfused (alveolar dead space ventilation V'_{dA}, Fig. 2.17b and
2. a compartment that is perfused (shunt perfusion Q'_S) but not ventilated.

As the "ideal" compartment is a theoretical concept (Fenn 1946), the ideal alveolar partial pressures are not measurable and the ideal compartment is not accessible. The problem can be solved with the aid of the respiratory quotient (R, equation 2.21) and mass balance in steady state, as described by equations 2.21, 2.22, and 2.23. The following steps lead to the alveolar gas equation (eq. 2.23):

- R of the ideal compartment must be equal to R of the entire lung.
- Since the slope of the CO_2 dissociation curve is very steep, it can be assumed that P_{aCO_2} is equal to P_{ACO_2} (Fenn 1946). This also holds true in the presence of venous admixture ($Q'_s/Q'_T > 0$, eq. 2.25), as long as Q'_S/Q'_T remains below 20% (Rossier 1960).
- The amount of nitrogen inspired is equal to the amount expired, i.e., no nitrogen is taken up by the blood.

$$R = V'_{CO_2}/V'_{O_2} \quad [2.21]$$
$$P_{ACO_2} = P_{aCO_2} \quad [2.22]$$
$$P_{AO_2} = F_{IO_2} \cdot (P_B - P_{H_2O}) - P_{ACO_2}/R + F_{IO_2} \cdot P_{ACO_2} \cdot (1 - R)/R \quad [2.23]$$

where P_{H_2O} is the water vapor pressure of the respiratory gas and P_B is the barometric pressure (in torr).

The partial pressures of the ideal compartment are now known. Since in this simplified model the inspiratory and expiratory gas movement (V'_E) is equal to the alveolar ventilation (V'_A), the dead space ventilation V_D/V_T can be calculated from V'_E (eq. 2.24). If the O_2 content of the mixed venous blood (C_{vO_2}) is known, the shunt function Q'_S/Q'_T can also be computed (eq. 2.25). The cardiac output (CO eq. 2.26) is calculated according to the Fick principle using measurements of O_2 uptake as well as O_2 content of arterial and mixed venous blood (avD_{O_2}):

$$V_D/V_T = 1 - V'_{CO_2}/V'_E/F_{ACO_2} \quad [2.24]$$
$$Q'_S/Q'_T = (C_{ACO_2} - C_{aCO_2})/(C_{AO_2} - C_{\bar{v}O_2}) \quad [2.25]$$
$$CO = V'_{O_2}/(C_{aO_2} - C_{\bar{v}O_2}) \quad [2.26]$$

with $F_{ACO_2} = P_{ACO_2}/P_B - P_{H_2O})$.

For reasons of accuracy, in practice the *mixed venous* O_2 saturation and not the partial pressure (steepness of curve, see Fig. 2.15) should be measured. Conversely, the *arterial partial pressure* should be measured and not the arterial O_2 saturation. P_{aO_2} can be converted into O_2 content with the aid of computer algorithms provided by Kelman (see list of formulas, appendix 8).

2.4.4. Analysis of the CO_2 Diagram

The Riley model has the major disadvantage that it contains the anatomical dead space in V_D/V_T. V'_E does not represent V'_A but rather $V'_A + V'_{dAW}$. The airway dead space determined by means of the N_2 washout (section 2.3.5) can be used to calculate V'_A. An easier way (without washout maneuver) is the single breath CO_2 diagram.

In analogy to the N_2 diagram, the CO_2 diagram is plotted as F_{CO_2} versus V_E. Mechanisms responsible for the generation of the CO_2 diagram are similar to those that bring about the N_2 diagram. During inspiration, interfaces are formed in the acini analogous to the interfaces formed during the nitrogen washout. The positions of the interfaces are determined by inspiratory flow, diffusion coefficients, and geometry. However, while the alveolar N_2 concentration depends on the relative ventilation,

(V_A/V_L, see Chapter 2.3.5), the CO_2 concentration depends on the ventilation/perfusion ratio.

2.4.4a Series Dead Space and Alveolar Ventilation

The airway dead space PIE_{CO_2}, for example, has an indefinitely high V'/Q' because it is not perfused. PIE_{CO_2} is determined using the same method as PIE_{N_2} (equations 2.10 to 2.12) and hence V_A and V'_A can be calculated in each breath n and without N_2 washout maneuvers (eq. 2.27).

$$V_A(n) = V_E(n) - PIE_{CO_2} \qquad [2.27]$$

2.4.4b Alveolar Efficiency of CO_2 Elimination

In the ideal lung, CO_2 flows from the blood into the alveolar space until the CO_2 partial pressure of the arterial blood has been reached. In this case one would expect an expired CO_2 volume, which is equal to the product of V'_A and F_{ACO_2}. In fact, however, only V'_{CO_2} is measured. The quotient derived from the amount of CO_2 expected and the amount actually measured is called the alveolar efficiency of CO_2 elimination (alv.eff.-CO_2) and can be determined *in steady state* according to eq. 2.28 for each breath provided that P_{aCO_2} does not vary.

$$\text{alv.eff.-}CO_2(n) = V_{CO_2}(n)/V_A(n)/F_{ACO_2} \qquad [2.28]$$

An uneven distribution of ventilation and perfusion too can lead to reduced alv.eff.-CO_2 or to Q'_S/Q'_T. Fig. 2.18 gives an illustration of a distribution abnormality with a

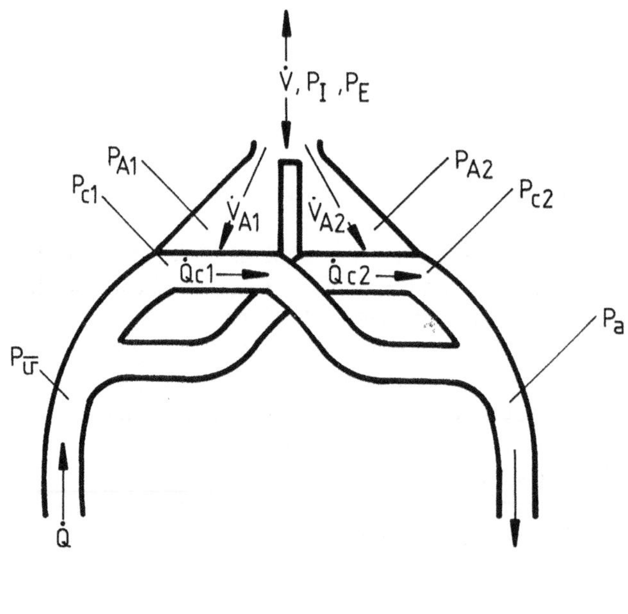

Fig. 2.18 Simplest model of inhomogenous V'_A/Q' distribution: two compartements with different ventilation-perfusion relationship ($V'_{A1}/Q'_{C1} \neq V'_{A2}/Q'_{C2}$). The greater the difference, the lower the efficiency of gas exchange. This is why inhomogeneity can be expressed as deadspace and shunt (venous admixture).

bimodal distribution with standard deviations of 0. The greater the difference between V'_{A_1}/Q'_{c1} and V'_{A_2}/Q'_{c_2}, the lower the efficiency of the gas exchange and the greater the dead space ventilation (see appendix 5, "Relationship between V'_A/Q' scatter and dead space").

Wagner and co-workers (1974) developed an excellent method to determine the different V'_A/Q' ratios present in the lungs. Like the Riley method, it is a steady state method and based on the Fick principle (eq. 2.20). It is applied to an ideal lung unit (see Fig. 2.17a), assuming the exchange of an inert gas with linear relationship between partial pressure and content. The principle of mass balance states that the expired amount of gas must be equal to the content difference between mixed venous and arterial blood in steady state (eq. 2.29).

$$P_A \cdot V'_A = C_{\bar{v}} \cdot Q' - C_A \cdot Q' = \lambda \cdot P_{\bar{v}} \cdot Q' - \lambda \cdot P_A \cdot Q' \qquad [2.29]$$

where the solubility of the gas λ is corrected for BTPS (partition coefficient). Rearranging this equation gives the definition for inert gas elimination as defined by Farhi, 1967 (eq. 2.30).

$$P_A/P_{\bar{v}} = \lambda/(\lambda + V'_A/Q') \qquad [2.30]$$

Since the ratio P_A/P_v depends only on λ and V'_A/Q', equation 2.30 indicates a method to investigate V'_A/Q' relationships. For an inert gas with a large partition coefficient, the ratio P_A/P_v is sensitive to high V'_A/Q's, whereas for an inert gas with a small partition coefficient it is sensitive to low V'_A/Q's. The investigation of a wide range of V'_A/Q' therefore requires the use of different inert gases which cover a wide range of partition coefficients.

The inert gas method of Wagner and co-workers (1974) uses this principle. Six gases of different solubility are dissolved in dextrose and infused at a constant rate into a peripheral vein. When steady state is reached, the mixed venous, the arterial, and the mixed expired partial pressures are measured using a gas chromatograph. The resulting ratios $P_a/P_{\bar{v}}$ (retention) and P_E/P_v (excretion) are plotted against solubility and form the basis for interpretation.

The final step is to combine the retention and excretion diagrams into one, the so-called V'_A/Q' diagram. This is done using a least squares fitting approach with enforced smoothing and assumes 50 lung compartments arranged in parallel with different V'_A/Q' ratios. The result gives the hypothetical amounts of ventilation and perfusion for each compartment (distribution) which best fit the measurements.

Although this method is a powerful tool to describe gas exchange, its result is not unique. Mismatch of ventilation and perfusion, pathological changes of diffusion capacity, intraalveolar mixing deficiency, as well as temporal changes in ventilation and within one breath can result in the same V'_A/Q' diagram (Hlastala 1981).

It should be noted that the difficulties in applying the method are not only associated with the computation of the V'_A/Q' diagram, but also with the preparation and infusion of the inert gas solution. Furthermore, the problems with gas chromatographic measurements should not be underestimated. Users of the method say that one study with three different experimental conditions keeps one experienced technician busy for 10½ hours (Hedenstierna 1987). Moreover, a principle critique to the method is that the ventilation of the anatomical dead space cannot be separated from the ventilation of the compartments with $V'_A/Q' = \infty$.

It might be valuable to point out that eq. 2.30 can be used to explain the limitations of the Riley model. In principle, the equation is only valid for inert gases. For the present purpose and for small partial pressure intervals, however, the dissociation curves of CO_2 and O_2 may be expressed as "partition coefficients". Used in this sense and because of the steep dissociation curve for CO_2, the "partition coefficient" of CO_2 is very high and therefore the ratio $P_A/P_{\bar{v}}$ is sensitive to high V'_A/Q' ratios. It is insensitive to low V'_A/Q' ratios, which forms the basis for using the arterial P_{CO2} as the ideal alveolar P_{CO2}. In other words, CO_2 is appropriate only to measure dead space ventilation.

The "partition coefficient" of oxygen, on the other hand, is high at low partial pressures and decreases towards higher partial pressures (S-shaped curve, see Fig. 2.15). This implies that P_A/P_v is sensitive to both low and high V'_A/Q''s depending on the partial pressure.

Finally, the Riley model uses only two gases. As a result, the V'_A/Q' resolution is limited.

2.4.4c Slope of the Alveolar Plateau

The CO_2 also provides information on V'_A/Q' scatter (West 1957). They are contained in the slope of the alveolar plateaus (phase III). The slope of phase III ($slope_{CO_2}$) is determined in two steps:

a) focusing on phase III. The portion of the CO_2 diagram between $PIE_{CO_2} + (V_E - PIE_{CO_2})/4$ and $PIE_{CO_2} + 3 \cdot (V_E - PIE_{CO_2})/4$ is analyzed;
b) approximation of the selected portion by means of a straight line.

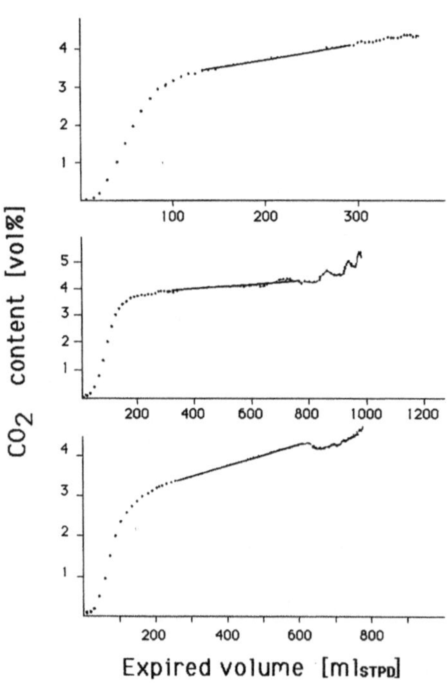

Fig. 2.19 Examplex of CO_2-diagrams measured in patients. Dots are measured samples (60 times per second). Solid line is approximation of alveolar plateau as described in the text.

The least squares fitting procedure is not a suitable method for this purpose, because the points are not evenly distributed. A mean value method is therefore used. The selected part of phase III is divided into two halves in respect to volume. The instantaneous samples (see part III) are averaged separately in both halves. The straight line through the mean values is the approximation of the alveolar plateau (examples in Fig. 2.19).

For the alveolar plateau to rise, the V'_A/Q' must be non-uniformly distributed, that is, there must be regional differences in CO_2 concentration. If these regions empty in sequence so that high V'_A/Q' first appear at the mouth, the alveolar plateau rises. The greater the V'_A/Q' scatter, the steeper the $slope_{CO_2}$. On the other hand, a horizontal phase III does not necessarily mean a homogenous V'_A/Q'. A particular pattern of emptying (synchronous emptying, for example), can effect a horizontal plateau despite V'_A/Q' inhomogeneities (masking). According to West (1957), however, the factors that cause V'_A/Q' inequality are probably also responsible for sequentiality.

There is another factor besides V'_A/Q' inhomogeneity that is responsible for a rising alveolar plateau. During expiration the lung volume continuously decreases. Since the lung perfusion and hence the diffusion of CO_2 into the alveoli continues or even increases, the alveolar CO_2 concentration during expiration also rises and produces a sloping alveolar plateau.

Nota Bene:

- The CO_2 single breath diagram contains information on the ventilation-to-perfusion relationships and on the emptying patterns of the lungs. The alveolar dead space ventilation must be regarded as inefficiency of the CO_2 elimination and must not be interpreted as actual dead space ventilation.
- The formation of the alveolar plateau is so complex that extreme caution is called for when interpreting its slope.
- Most of the indices described in this section are based on Riley's three-compartment model, which has been used for quite some time. The relatively tedious calculation work required for determining these indices makes the model impractical for clinical practice. This book and the software made available will hopefully be a contribution to overcoming this barrier.

3. Assessment of Pulmonary Function Indices

3.1. Sensors

3.1.1. Measuring Head

The gas flow, airway pressure, and gas composition are measured directly at the mouth, i.e., between the respirator and the patient, in the so-called "measuring head" (Fig. 1.1). This consists of a Fleisch pneumotachograph (PT) No. 2 and two Teflon adapters, which provide a "soft" transition from the narrow endotracheal tube and the ventilator connector to the diameter of the PT. A catheter to measure the airway pressure (perpendicular to the flow) and the capillary for the mass spectrometer (tip in the center of the gas flow) are inserted into the Teflon adapter which is closer to the patient.

3.1.2. Flow Measurement

3.1.2a Principle of Measurement

The flow of gas is measured with a resistive flow transducer (Fleisch 1925). This measuring technique can be applied in patients with little secretion and who move only very little. Since it is not dangerous, is easy to apply, and can be sterilized, it is highly suitable for clinical application. However, it yields results that have an error of 10–20%, which is caused by changes in gas viscosity, temperature, and upstream geometry. The physical law on which the Fleisch flow transducer is based was formulated by Hagen and Poiseuille. It states that the relation between volume flow dV/dt through a long rigid tube with a diameter of 2r and the pressure drop Δp across l is:

$$dV/dt = \eta \cdot r^4/(8 \cdot \pi \cdot l) \cdot \Delta p \qquad [3.1]$$

where η = dynamic viscosity of the fluid and l the distance over which Δp is measured. Equation 3.1 applies only under the following conditions:

- steady state
- incompressible gas
- laminar flow
- completely developed boundary layer.

Before we can discuss to what extent the *Fleisch Pneumotachograph* (PT No. 2) meets these conditions, it is necessary to describe its construction in detail.

The inner diameter of the PT No. 2 measures 28 mm and is completely filled with small parallel tubes having an inner diameter of $D_T = 0.96$ mm and a length of $l_T = 29.4$ mm. Assuming that the gas flow is the same in each tube, it does not matter which of the approximately 850 tubes is used for the flow measurement. For practical reasons, measurement is done in a tube located at the periphery so the pressure can be measured at two sites perpendicular to the direction of the flow. The measuring sites are located 4.7 mm from the end of the respective tube and about 20 mm away from one another. The pressure difference Δp is fed to a pressure difference transducer (Statham PM 15 TC) via gas-filled tubes (electrical isolation). An amplifier (Analog Devices, Type 3B18) with a bandwidth of 1 kHz and an integrated bridge supply is used to measure the small signal from the pressure difference transducer.

Since the tubes are very long in comparison with their diameter ($l_T/D_T = 30.6$), the pressure drop as a result of friction (resistive effect) is greater than the pressure drop caused by mass acceleration. Therefore, for all practical purposes, the *steady state* requirement can be considered fulfilled, although the PT is also used for measuring flow changes, i. e., non-stationary flow.

The pressure gradient compresses the gas volume. This is so slight, however (only a few mmH_2O), that for practical purposes the gas can be assumed to be *incompressible.*

The last two conditions – laminar flow and completely developed boundary layer – can be examined using the Reynold's number Re (eq. 3.2), which indicates whether flow in a tube with a diameter D_T is more likely to be laminar or turbulent.

$$Re = v_m \cdot D_T/\mu \qquad [3.2]$$

Re and hence the transition from laminar ($Re < 2300$) to turbulent ($Re > 2300$) flow depends on the mean velocity v_m as well as on the kinetic viscosity μ of the flowing gas. According to equation 3.2, this transition takes place in the tubes of PT No. 2 at approximately 49 m/sec ($\mu_{air} = 2.05\ 10^{-5}\ m^2/sec$ at 720 torr and 37°C). The flow in the tube thus remains *laminar* up to a flow of 35 ml/sec. In order to establish a completely developed boundary layer (Poiseuille flow), a minimal entrance length l_e must be available.

$$l_e = 0.03 \cdot D_T \cdot Re \qquad [3.3]$$

For the PT No. 2, the maximum entrance length is the distance of the measuring sites from the end of the tube (4.7 mm). The maximum admissable Reynolds' number (solution of eq. 3.3 for Re) is thus 163, i. e. the maximum velocity in the tube is limited to 3.47 m/sec (v_m from eq. 3.2). The range of the measurable volume flows computed from this is 2.48 ml/sec per tube. For the PT No. 2 this number must be multiplied by 850 tubes, which yields a theoretically predicted working range of ± 2.1 l/sec.

Up to this point it has been assumed that the flow is identical in all of the tubes. This is true as long as the velocity profile is flat before it enters the PT. If it changes, the relationship between flow and Δp in the PT also changes. Sharp edges, valves, and tubing diameter changes with beam angles that are too large lead to unforeseeable changes in the velocity profile and hence to errors in the flow measurement.

Our set up (see Fig. 1.1) is a compromise between the demands of the theory of fluid dynamics and those made by the clinic. The beam angle of the Teflon adapter is

10° (Becker 1977). We retained the angled connector to the tube ("swivel") commonly used in our intensive care unit. The swivel does not introduce a flat velocity profile, but it at least keeps parabolic velocity profiles from occurring upstream to the PT. It also makes it possible to perform tracheobronchial suctioning without having to disconnect the measuring head (Wolff 1983, page 178ff.). We made no attempt to influence the inspiratory flow pattern using constructional means. A simple possibility, for example, would be to mount a commercially available "linearizer" in front of the PT to try to force a flat velocity profile. This, however, would enlarge the apparatus dead space.

3.1.2b Linearity

The linearity of the flow measurement in the arrangement shown in Fig. 1.1 was investigated. By "linearity" is meant the proportionality of the volume flow to the pressure difference measured on the PT (measured here in units of A/D units ADU). To emphasize deviations from the linearity, the first derivative is considered in dependence on ADU (so-called conductance curve). A convenient way to calibrate a flowmeter is to take a constant source of flow with an adjustable flow rate. A syringe could be used with a piston moved at a constant speed. This, however, is a difficult and costly engineering task which, in fact, is not even necessary. The curves of the PT can be calculated with 1% accuracy from 50 to 100 strokes by hand with a syringe of an exactly known volume. The following is a brief description of the calibration method devised by Yeh (1982).

The zero-line is assumed to be known. It is further assumed that one single flow-rate can be assigned to each instantaneous value of the A/D transducer (ADU). In the first stroke all of the instantaneous values are added up until the volume of the syringe (one liter) has passed through the PT. This sum, when multiplied by the calibration factor and the sampling interval (here 16.66 ms), must be exactly 1 liter. In theory, the calibration factor of a particular flow can be calculated from this under the assumption that the syringe was pushed at a constant speed.

Since calibration is done by hand, this assumption is not exactly accurate and there will be a somewhat broad distribution of flow. For this reason we count how often each individual instantaneous value comes up during one stroke. The calibration factor determined is assigned to each ADU and is weighted by how often this instantaneous value appeared.

There is no sense using this procedure for one single stroke, since only one calibration factor is computed. Further strokes are necessary to calibrate the entire flow range. Each stroke of different flow rate gives a new calibration factor. Each stroke of different flow rate also produces different ADU's. The calibration factors are assigned to the ADU's according to their frequency (weighted mean values), which yields the first approximation of the calibration curve after a sufficient number of strokes (here 84). This calibration curve is called "initial calibration curve" and is needed for a second pass.

In the second pass, the corrections of the initial calibration curve are computed analogously to the procedure described above. Using the initial calibration curve, the predicted volume is calculated and the relative deviation from the expected value

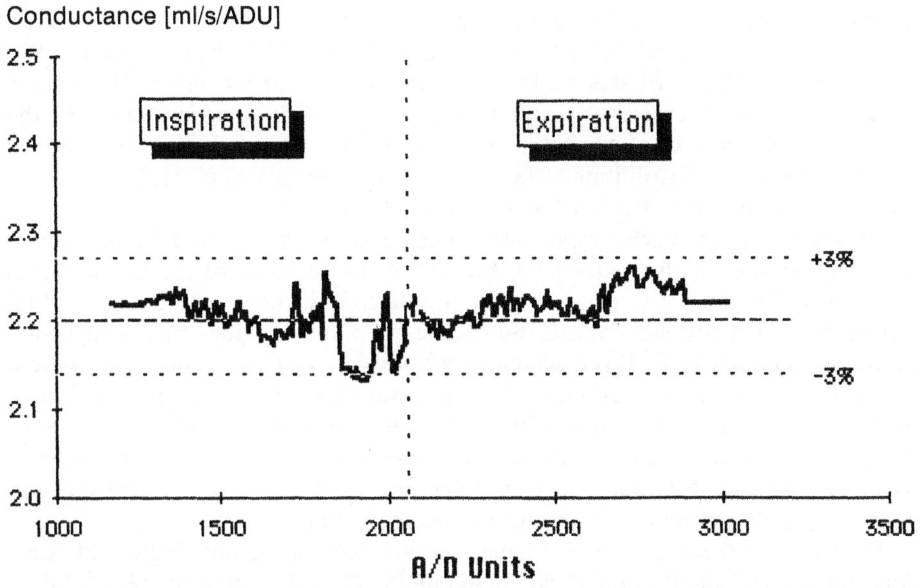

Fig. 3.1. Calibration curve of the PT in the set up according to Fig. 1.1. The A/D units plotted on the abscissa are proportional to the differential pressure over the PT. The vertical dashed line marks the zero line. The horizontal dashed lines represent the mean conductance and the ± 3% limits

(1 liter) is determined. This error is then assigned to the ADU's that came up during the stroke. The weighted mean error is determined for each ADU from several strokes. A second approximation of the calibration curve is produced by correcting the calibration factor for each ADU according to its mean error.

A third and a fourth pass were performed using the result of the second and third pass, repectively, and the result of this is represented graphically in Fig. 3.1.

The *results* show that the calibration curve of the PT is asymmetric, whereas ideally it should be straight and horizontal. The expiratory calibration curve is closer to the ideal than the inspiratory. The error, however, remains within ± 3.5%. We consider *one* mean calibration factor for all of the ADU's to be sufficient and therefore allow for a maximum error of ± 3.5%.

3.1.2c Calibration

In calibrating the flow, it is assumed that the proportionality factor between pressure decrease and flow remains constant (absolute linearity, see above). The integration of the flow to the volume is then independent of the calibration factor, since this remains constant.

Using a precision air-syringe, 1 liter each of expired and inspired air is passed by hand through the PT and the instantaneous values measured during this time minus baseline are added up separately for inspiration and expiration. One liter divided by the corresponding sums yields the calibration factors for inspiration and expiration (variable KVOLIN and KVOLEX in Fortran Program SCALER, see appendix 8).

The reproducibility of this calibration was investigated by passing 40 volume strokes by hand (1 liter ambient air with a measured precision syringe) through the pneumotachograph. An attempt was made to produce a ramp-shaped flow pattern, i. e., a uniform flow distribution. The evaluation yielded stable values, the scatter of the calibration factor was ± 0.6% (variation coefficient).

The Fleisch pneumotachograph must be heated to keep water from the ambient air from condensing on the wall of the tube. The temperature of the transducer is determined by the ratio of energy input (electric heater) and energy output (convection, heat conduction, radiation). The temperature under measuring conditions for patients is 32°C. It is much higher prior to the measurements, since there is no gas flow through the transducer. The pneumotachograph is therefore calibrated without being heated. The change in temperature during the measurements on the patient are compensated for mathematically under the assumption that the inspiratory gas heats up 5°C and the expiratory gas cools down 5°C (results of our measurements with rapid thermosensor in two patients).

The absolute accuracy of the volume measurements using one single calibration constant was measured in the range of 100 ml to 1000 ml in steps of 100 ml. Seven volume strokes per step were pushed by hand through the PT.

Results: there was a better than 3% agreement between the values measured from the PT and the expected values (Fig. 3.2 top). The scatter was ± 10 ml, of which approximately half could be attributed to the imprecision of the manual operation of the syringe, as the 100 ml steps of the 1 liter syringe were only marked with tape. The regression analysis yielded a slope of 0.978 at an r^2 of 0.9998.

3.1.2d Correction of Errors Caused by Viscosity Changes

When the viscosity of the gas flowing through the PT changes, the relationship between the pressure difference and the volume flow also changes (Fig. 3.2). To achieve flow measurement accuracy of better than 20%, the viscosity of the gas mixture must be taken into account. It is hypothesized that a linear approach is sufficient (see appendix 6, "Viscosity of gas mixtures"). This was tested using 12 different argon mixtures which were pushed through the pneumotachograph by means of a 700 ml syringe. All values were related to the value obtained with air and are expressed as percentages. Fig. 3.2 bottom shows the results in dependence on the argon concentration. The dotted line represents the relationship which would be expected on the basis of theoretical calculations (see appendix 7.6). The slope of this line is determined by the difference between N_2 viscosity η_{N_2} and argon viscosity η_{Ar} according to eq. 3.4. The volume V for air was set to 100%.

$$\Delta V / \Delta F_A = 100\% \cdot [\eta_{A_r} - \eta_{N_2}] / [0.2 \cdot \eta_{O_2} + 0.8 \cdot \eta_{N_2}] \qquad [3.4]$$

The theoretical value of $\Delta V / \Delta F_{Ar}$ calculated with the values in the appendix is 0.258, which means better than 2% agreement with the values in Fig. 3.2 (slope of the regression line is 0.262, $r^2 = 0.997$ is not plotted).

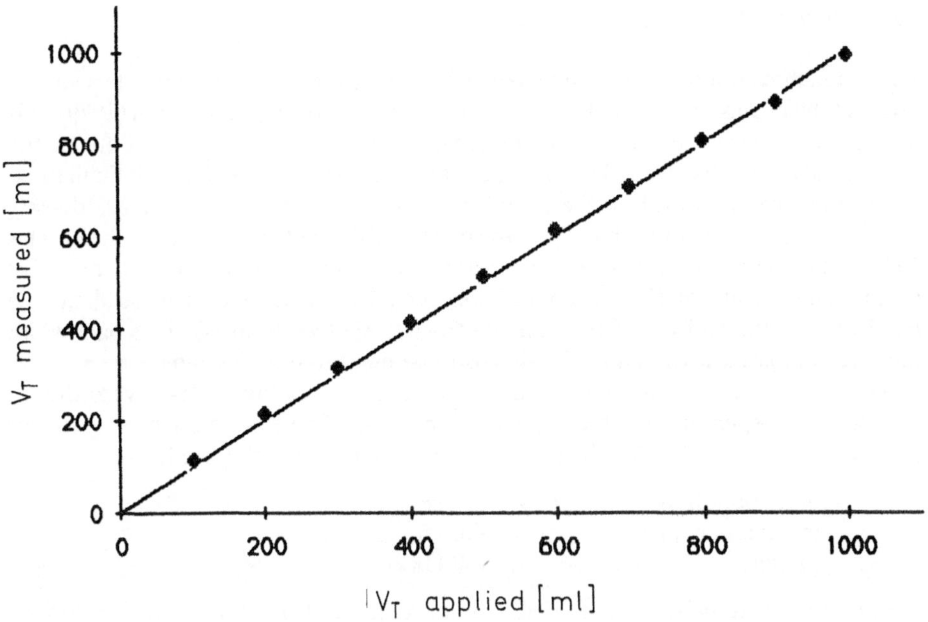

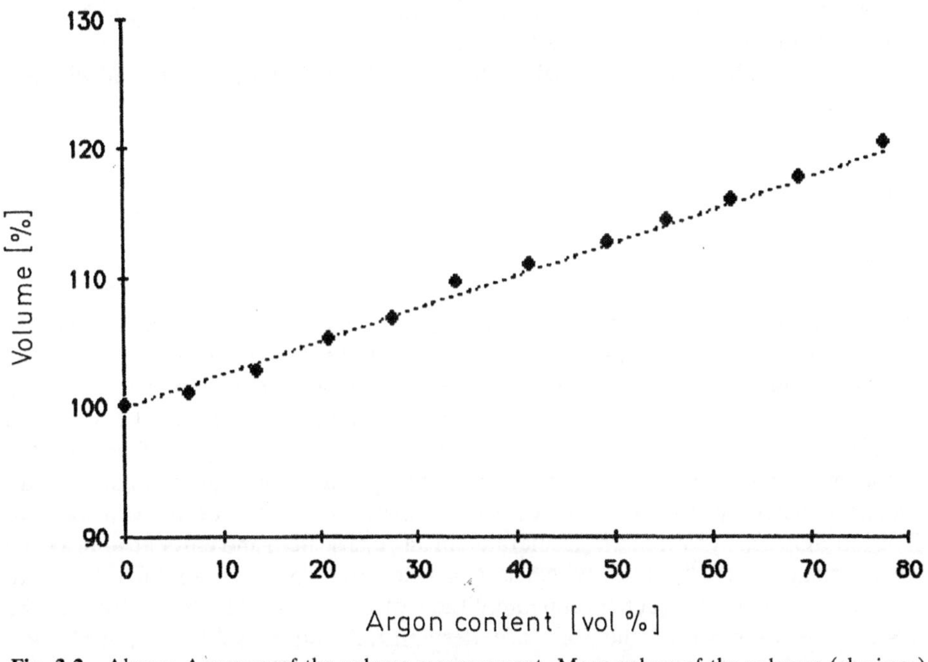

Fig. 3.2. Above: Accuracy of the volume measurement. Mean values of the volumes (abscissas) passed through the PT by hand using a syringe are plotted against the volumes measured from the PT. The solid line is the identity line. Below: Dependence of volume measurement (ordinate, volumes in % of the baseline value at 78 vol% N_2/21 vol% O_2) on argon content when there is a constant O_2 content (21 vol% O_2). The theoretically expected dependence (see also appendix 7.6) is represented as a dotted line

3.1.2e Bandwidth

The pneumotachograph is connected with the pressure difference transducer PM15TC by means of gas-filled tubes (see Fig. 1.1). According to Jackson (1979), it is the pressure transducer and the connecting tubes that limit the bandwidth of the system and not the PT itself. The longer the tubes are and the smaller their diameters, the smaller the bandwidth of the transmitted signals. Furthermore, the gas density also affects the bandwidth: the higher the density, the smaller the bandwidth (Francis 1979). The connecting tubes used here are 140 cm long with an inner diameter of 6 mm. The density of N_2 and Ar is lower than 1.8 g/l. Based on these data, the bandwidth of the entire system is smaller than 25 Hz (see Francis). This bandwidth must be compared with that of the signals to be measured in the patient.

The volume flow signals measured at the mouth of healthy adults have different bandwidths, depending on the type of respiration. In 95% of all of the signals measured by McCall (1957), the following bandwidths were reported:

quiet breathing	< 3.5 Hz
maximum expiration	< 4.0 Hz
panting	< 26.0 Hz

A more recent study by Lemen (1982) contains the spectrum analysis of forced vital capacity maneuvers. The volunteers inspired maximally, then took an end-inspiratory break, and expired maximally through the pneumotachograph using a mouth piece. The bandwidth of these flow signals was less than 12 Hz.

The bandwidth of the flow signals is thus < 30 Hz and can be determined with the measuring head.

3.1.3. Gas Analysis

3.1.3a Measuring Principle

A mass spectrometer (MS) is used to analyze the respiratory gas (Centronic MGA-200). With this device several gases can be measured simultaneously.

The functional principle can be described as follows: samples of the gas to be analyzed are continuously drawn into the inlet system (10^{-1} torr) of the MS via a capillary (electrical isolation). By means of a molecular leak, the gas diffuses into the actual analyzing chamber (10^{-7} torr). There the gas molecules are bombarded with electrons, whereby some of the molecules are ionized, ionized twice, or even cracked. The ions that are produced are accelerated in an electric field and enter an alternating electromagnetic (EM) field, which is generated by 4 parallel conductors (quadrupole). With the frequency and amplitude of the EM-field kept constant, the trajectories of most of the ions are destabilized and these ionized molecules consequently are trapped somewhere on the vessel wall. There they are neutralized and drawn off with vacuum pumps. Only ions with a particular mass-charge ratio have a stable trajectory. They pass through the alternating field and hit the photomultiplier. The photomultiplier thus generates a voltage which is proportional to the partial pressure. If the frequency or amplitude of the EM-field is changed, the ions with another mass-charge ratio reach the photomultiplier. In the MGA-200 the amplitude of the EM-field is changed in such a way that the ions with mass-charge ratios 28 (N_2), 32 (O_2), 44 (CO_2), and 40 (Ar) can be measured. The Centronic MGA-200 is designed in such a way that the EM-field change takes place 4 times per 20 milliseconds. Consequently, the time course of the partial pressure of all 4 gases can be measured simultaneously with a sampling frequency of 50 Hz.

3.1.3b Automatic Sensitivity Control

A control circuit in the MS keeps the sum of the partial pressures of CO_2, O_2, N_2, and Ar constant. The advantage of such a control is that undesired effects such as pressure changes at the measuring site, water vapor content of the gas, and aging of the photomultiplier can be compensated for. The measuring values therefore represent volume portions of dry gas (vol%).

3.1.3c Linearity and "Cracking Pattern"

Hallbäck (1977) studied the accuracy and the scatter of the CO_2 measurement with the Centronic MGA-200 using the Haldane technique as a standard. Ninety-nine samples of CO_2 in the range of 0 to 8.1 vol% were compared. The mean difference ± SD was 0.001 ± 0.045 vol%. According to Lerou (1984), the linearity for O_2, N_2, and Ar is better than 0.3 vol% between 0 and 100 vol%.

When the gas molecules in the ionization chamber are bombarded, a portion of the CO_2-molecules are cracked into CO-ions. CO-ions have the same mass-charge number as N_2-ions and therefore the mass spectrometer cannot differentiate between them. According to the manufacturer (Centronic), about 10% of the CO_2 is measured as N_2. This effect must be compensated for and is done so here (see appendix).

3.1.3d Calibration

The mass spectrometer is calibrated with air and a special gas mixture. This mixture consists of 7.8 vol% CO_2, 32.4 vol% Ar, 59.8 vol% O_2, and 0 vol% N_2. Since, as we said above, approximately 10% of the CO_2 appears on the N_2 channel, the mass spectrometer has to be calibrated as shown in Table 3.1. Only 1-point calibration is

Table 3.1. Gas mixture for calibration of the MS

Gas	Special Gas mixture	± 0.1 vol%	Air ± 0.1 vol%
CO_2	7.8 − 0.78 vol% =	7.0 vol%	0.1 vol%
O_2		59.8 vol%	20.9 vol%
N_2	0.0 + 0.78 vol% =	0.8 vol%	77.9 vol%
Ar		32.4 vol%	1.0 vol%
		100.0 vol%	99.9 vol%

possible so that the gas with the higher content (e.g. air for N_2) is used. The content of the other gas is used as a control and, if necessary, for any readjustments. Both gases, however, are used (2-point calibration) to scale the entire measuring chain from the mass spectrometer through the amplifier, filter, and A/D transducer to the computer.

3.1.3e Transport Delay

For gas to be analyzed in the mass spectrometer, it has to be transported from the measuring site to the analyzing chamber via a capillary tube. This mass transport takes time and the results are thus delayed (delay time). Since specific gas flow (e.g. N_2 gas flow) is to be calculated later on by multiplying the concentration by the volume flow, the signals from the mass spectrometer have to be synchronized with those from the pneumotachograph. For this it is necessary to know the delay time. The specifications of the capillary tube are not exact, which means that delay time cannot be calculated, but must be measured for each capillary tube (see appendix 7, "Determination and dependencies of the delay time", Brunner 1985c).

To determine the delay time, a special syringe is used which produces a change in the gas composition at the entrance to the capillary tube. This change takes place at the same time a known volume is pushed out of the syringe. The time from the ejection of this volume to when the gas change is displayed on the MS is the delay time (details in appendix). The accuracy of the method is ± 10 ms.

During a washout maneuver, the gas viscosity changes from breath to breath. This in turn alters the delay time of the MS and can cause considerable errors (Brunner 1985c, see appendix). For this reason, breath-by-breath correction of the delay time is absolutely mandatory.

3.1.3f Response Time

If the concentration at the entrance to the capillary tube is changed in the form of an ideal step (Heavyside function), the measured response of the MS takes on the shape of a ramp. This is due to the limited response time of the MS, which is described by the time in which an increase from 10% to 90% of the end value takes place. The response time is primarily determined by the construction of the inlet system and in the MGA-200 amounts to approximately 80 msec. That means that when there is a maximum flow of 1 l/sec, a gas front with a dispersion of 80 ml (e.g. phase II) can just about be determined before there is any distortion.

3.1.4. Pressure Measurement

3.1.4a Measuring Principle

The airway pressure and esophageal pressure are measured with a P23Db Stratham pressure transducer. The measuring site is connected with the sensor via an air-filled catheter (electrical isolation). The pressure to be measured causes a membrane to bend, thereby altering the resistance of a strain gauge. An electronic circuit (3B18, Analog Devices) supplies this strain gauge and amplifies the differential signal (bandwidth 1 kHz).

3.1.4b Linearity

The transducers used here were originally used for blood pressure measurements and therefore operate in a range of − 68 to + 4080 cmH_2O. In our application only 2% of the total operating range is used (− 30 cmH_2O to + 50 cmH_2O). This range is linearized. With a water column in a U-tube for reference, accuracy and linearity were measured in static operation. The pressure was varied 19 times in steps of 5 cmH_2O and compared with the measured values. To expose any possible hysteresis effects, the pressure was first raised from − 20 cmH_2O to 30 cmH_2O and afterwards lowered again to − 20 cmH_2O. The pressure of the water column could be read with an accuracy of approximately ± 0.1 cmH_2O. The results are shown in Fig. 3.3 together with the identity line. The electrically measured pressures deviated from the reference values by 0.9% on average (slope of the regression lines was 1.006 with $r^2 = 0.9999$).

3.1.4c Bandwidth

The sensors used here have a stiff membrane and thus a large bandwidth (> 55 Hz). According to Francis (1979), the bandwidth decreases the longer and narrower the connector to the transducer is. To measure the airway pressure, a plastic connector (disposable article) having a length of 150 cm and a diameter of 1.5 mm is used. For the esophageal pressure measurement, an additional catheter of 123 cm in length and 1 mm in diameter is required with a deflated balloon attached at the end (see Fig. 1.1). Under these conditions, the bandwidth for the airway pressure measurement is greater than 30 Hz and for the esophageal pressure greater than 20 Hz (according to Francis). The latex balloon has no effect on the bandwidth as long as it is not fully inflated.

The maximum bandwidth of the pressure signals is the same as that of the flow signals and amounts to 30 Hz. The transducers are sufficient to assess the frequency range of all of the pressure signals.

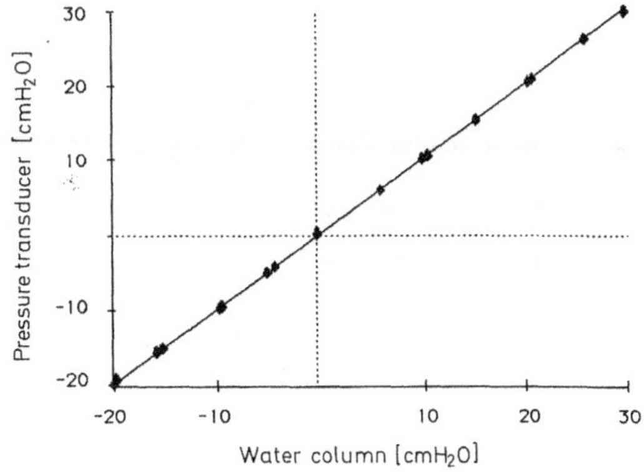

Fig. 3.3. Linearity of the statistic pressure measurement. The pressure of the water column is plotted on the abscissa. The values measured with the transducer are plotted on the ordinate

3.1.5. STPD and BTPS Conditions

Dry gas fractions can only be multiplied by dry gas volumes. For this purpose, all of the measured gas volumes are converted to STPD conditions (standard temperature, 273°K, standard pressure, 760 torr, dry gas). To do this, the temperature, pressure, and humidity of the measured gas have to be known. The temperature and humidity were not measured continuously to avoid having to use too many additional sensors when performing measurements on the patient. For the calculations, the following assumptions were made on the basis of our own studies and those of other authors (see Brunner 1983):

- the expirate cools down 5°C on the way from the lung to the measuring site
- the inspirate heats up 5°C in the pneumotachograph
- 80% relative humidity of the expirate
- 60% relative humidity of the inspirate.

The water vapor pressure P_{H_2O} in torr can be calculated with the aid of the body temperature or the ambient temperature T (in °K) according to eq. 3.5. The water vapor content results from this under consideration of the relative humidity:

$$P_{H_2O} = 13.2 - 0.61 \cdot (T\text{-}273) + 0.04 \cdot (T\text{-}273)^2 \qquad [3.5]$$

$$V_{STPD} = V_{BTPS} \cdot 273/760 \cdot (P_B - P_{H_2O})/T \qquad [3.6]$$

Conversion of the volumes is carried out according to eq. 3.6. Conversely, in calculating the distensibility (see Chapter 2) the distension of the gases in the lung must be taken into consideration, i.e., the volumes must be converted to BTPS (body temperature, body pressure, fully saturated) conditions. A correction of the respiratory pressure was not made, since its effect (mean pressure < 11 cmH_2O) amounts to less than 1%.

Units: The gases are heated to body temperature in the lung and are saturated with water vapor. Single breath volume, respiratory minute volume, compliances, and flow resistance are given in BTPS units (body temperature, body pressure, fully saturated). The specific gas volumina and the quantities derived from them are given in STPD units (standard temperature 273° Kelvin, standard pressure 760 torr, dry gas). Caution: Be aware of the difference. Under clinical conditions, it can be up to 25%. The gas compositions are given as dry gas fractions, i.e. in vol %.

Nota Bene:

- All sensors must be placed at exactly the same site in the gas stream, i.e., directly at the mouth.
- Flow and pressure sensors should have a bandwidth of at least 25 Hz and an accuracy of better than 5%. The bandwidth decreases with increasing length of the connector from measuring site to sensor.
- Principally, direct measurement of the flow cannot be precise. The flow signal has to be continuously corrected with respect to the viscosity, water vapor content, and temperature.
- Gas analyzers should have an accuracy of better than 1% (relative).

- Direct concentration measurements are not precise. If a mass spectrometer is used, the water vapor problem must be eliminated by automatic sensitivity control and the problem of the cracking pattern solved by correcting the concentration measurements in each sample.
- The delay of the mass spectrometer with respect to the flow and pressure signals (delay time) is dependent on the composition of the respiratory gas. Therefore, the delay time is not constant and must continuously be adjusted.
- Whether the response time (from 10 to 90% of a step) of the gas analyzer is short enough depends on the expiratory gas flow. If the response time is greater than 80 ms, an irreversible loss of information can occur during measurement in ventilated patients.

3.2. Data Processing

3.2.1. Hardware and Data Flow

All of the programs were developed on a DEC computer LSI-11/23 with Memory Management (RAM up to 256 kilobytes, available storage space 196 kilobytes). A floating point processor (FPP) provides quick multiplication and division (< 0.08 ms for single precision). Peripheral equipment include a printer (Star Radix 10), a double floppy disk station (RXO2, 2×500 kilobytes), and a graphics video terminal (VT 100, Retrographics or Apple Macintosh with suitable software). Analog data are acquired via the 16 channels of an A/D converter (< 0.05 ms conversion time per channel). The system timer (Line Time Clock LTC, 60 Hz) is used to control the conversion.

Analog to digital (A to D) conversion is critically dependent on the rate at which analog values are converted to digital numbers, i.e., the sample rate. Fortunately, there is a thorough mathematical theory for this problem, which finds its simplest yet powerful expression in the Shannon theorem. This states the sampling frequency must be at least twice the highest spectral component of the analog signal to be converted. Here, the 60 Hz sample rate allows a maximum signal bandwidth of 30 Hz. If higher frequencies occur, unrecoverable distortions result. This effect is called aliasing and is due to convolution effects. Therefore, a critically damped low pass filter of a 2nd order with a roll off frequency of 25 Hz was used to prevent aliasing. However, since in most cases the entire spectrum of a signal is to be assessed, the use of a low pass filter has a merely preventive character. Its purpose is to filter out high frequency noise.

Fig. 3.4 gives a schematic representation of the data flow. MAIN stands for all of the programs that require calibrated, corrected, and synchronized raw data, as, for example, the program called MC to determine C_L, C_{tot}, R_{AW}, etc.

MAIN obtains a set of raw data, i.e., an instantaneous value of V', F_{N_2}, F_{CO_2}, F_{O_2}, P_{awo}, and P_{es} by calling the subroutine INVAL. Depending on the configuration of INVAL, the data are read from the disk or from the buffer "BLK1". They are then calibrated, corrected, and synchronized, and written into buffer "VALUES".

When the data are read from the disk, no time problems arise if the program MAIN needs a long time (seconds or minutes) for processing. This is not the case, however, with direct access to the buffer "BLK1", since this is written 60 times a second with values from the A/D converter. Reading data from the A/D converter and writing it into "BLK1" are tasks that must be performed synchronously with the system timer. They are therefore taken over by a so-called interrupt-service routine (ADCONV).

ADCONV is initialized from the MAIN program, which sees to it that an interrupt from the system timer (60 times a second) invokes a call to ADCONV. ADCONV then fills the buffer "BLK1" with values from the A/D converter, which results in a data-rate of $60 \cdot 6 = 360$ values or 720 bytes per second. "BLK1" can accomodate only 3360 values and is therefore full after 4.6 seconds. INVAL thus has the task of emptying the memory storage as fast as possible to make room for new data. It carries out this task by removing a set of data from "BLK1" each time it is called from within MAIN calibrates these values, corrects and synchronizes them, and then writes them into the buffer "VALUES".

It is the job of the main program MAIN to call INVAL often enough, i.e., an average of 60 times per second, to prevent the overwriting of yet unread data ("buffer overrun"). INVAL checks for buffer overrun and management of the buffer. Any error messages are forwarded to MAIN.

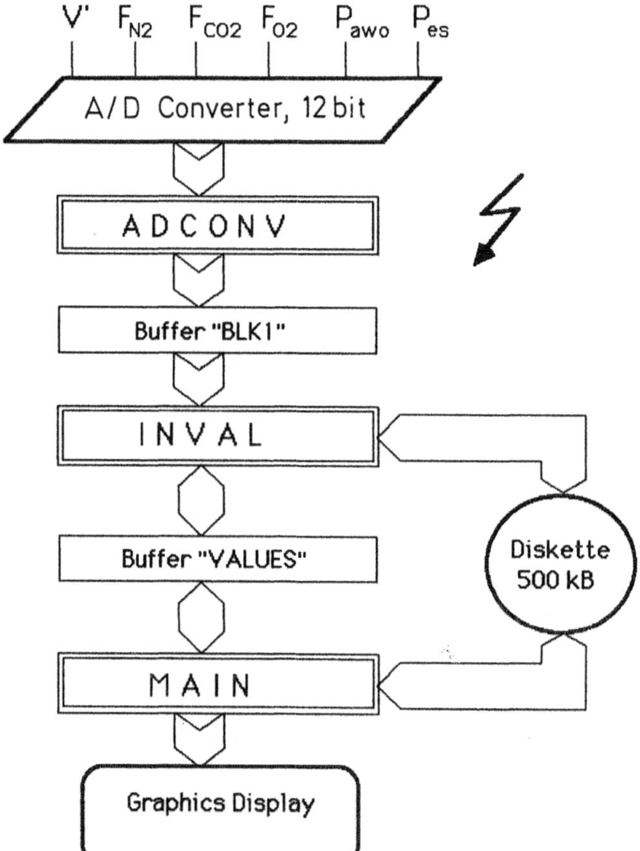

Fig. 3.4. Data flow diagram. The interrupt service routine ADCONV transports a set of instantaneous values into the BLK1 (buffer BLK1) 60 times a second. At each call, the subroutine INVAL checks the status of buffer BLK1 and empties when it is full onto a disk. INVAL synchronizes and corrects a set of instantaneous values (raw data) and writes it along with any error messages into the buffer "VALUES". Every main program that requires raw data (MAIN) must call INVAL, after which it can pick up a set of instantaneous values in buffer "VALUES"

The advantages of this buffered raw data acquisition are that MAIN does not have to call INVAL at fixed times intervals. This simplifies the programming considerably and provides for simple algorithms. Furthermore, the structure of the main programs is not dependent on whether the raw data are read from the A/D converter or from the disk.

3.2.2. Automatic Recognition of the Respiratory Phases

One respiratory cycle consists of the phases inspiration and expiration. The ability to recognize individual breaths with certainty and to be able to mark them off against other breaths is the basis of an automatic analysis. Zero point drift, coughing, swallowing, and cardiogenic oscillations hinder automatic recognition of the breathing cycles.

The algorithm used here accounts for these difficulties. It is based on the combination of flow and CO_2-concentration and its application in patients has shown it to be extremely robust (Brunner 1985b).

a) The *beginning of an expiration* (E) is marked by the last zero crossing of the flow before there is an increase in the CO_2-concentration at the airway opening (Fig. 3.5).

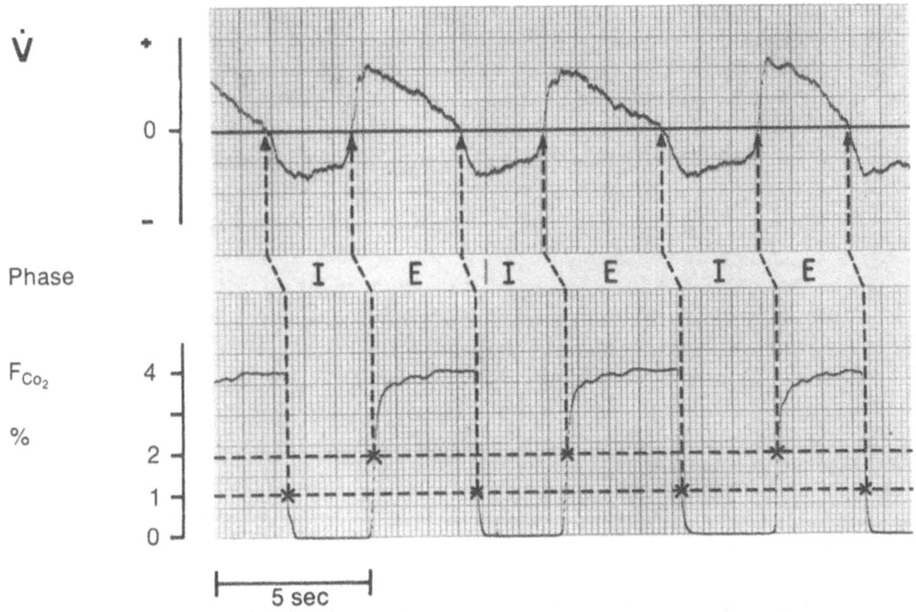

Fig. 3.5. Gas flow (V′) and CO_2 content (F_{CO_2} in vol%) as function of time. Each zero crossing of the flow (arrows) is a potential phase change. Two threshold values (horizontal dashed lines) exist for CO_2. If 2 vol% is exceeded (x) the previous flow zero crossing is confirmed as the beginning of expiration (E) (vertical dashed line). If CO_2 drops to below 1% (x) this is assumed to be the beginning of inspiration (I) (vertical dashed line)

b) The *beginning of an inspiration* (I) is marked by the last zero crossing before there is a decrease of the CO_2-concentration at the airway opening.
c) The *end of a phase* is identical with the beginning of the other phase.

Using these definitions, a simple algorithm can be set up as the basis of all the breath-by-breath calculations. Note that "CO_2 increase" means exceeding 2 vol% and "CO_2 decrease" means a drop to below 1 vol%. The separation of a flow pattern into inspiration and expiration is illustrated in Fig. 3.5.

3.2.3. Software

Data acquisition is done at a sample rate of 60 Hz (twice the bandwidth of 30 Hz). The data is passed to INVAL and optionally stored on disks without being altered i.e., without correction, calibration, and synchronization. Data acquisition takes place independent of the data analysis. As a result, it is possible to use any transducer without changing the software. This also allows the calibration constants to be determined as often as desired, simply and under visual control, and to be stored separately (two-point calibration). Two kinds of main programs have been written: programs that analyze raw data stored on disk and programs that analyze raw data from the A/D converter (real time programs). An application of the real time analysis is shown in Fig. 3.13.

Nota Bene:

- The computer hardware must be engineered in such a way that the maximum data flow rate can be easily managed. Since most modern computers fulfill this requirement, highest priority is put on the software.
- By using the flow signal combined with the CO_2 signal, absolute differentiation of the two respiratory phases (beginning of the inspiration phase, beginning of the expiration phase) can be achieved with exact timing and no error.
- The software was developed in the Department of Physiology in Basel according to the local circumstances. Therefore, there are limits to its general use. Nevertheless, we would advise the interested programmer not to go through reinventing the wheel. We can provide the entire software package (Source Code in FORTRAN IV and Macro-11 under RT-11) including detailed documentation (4 disks 8", RT-11 format) for the cost of the copying and mailing expenses (about 100 Swiss francs).

3.3. Testing the Measuring System: Results and Discussion

Simple means cannot be used to test absolute accuracy of the measuring system when used in patients. One reason is the inaccessability of many quantities (e.g. FRC) and another is that several indices represent concepts and not directly measurable quantities. These problems led to the development of the physical lung model described below, with which the absolute accuracy of several indices can be determined.

In many cases, it is not the absolute quantities that are important for the clinical assessment but rather their changes. For this reason, the reproducibility of the measurements in patients were investigated.

3.3.1. Absolute Accuracy

3.3.1a Physical Model of the Lung

The purpose of the physical model of the lung introduced here is to test non-invasive measurements at the mouth of a patient. The following lung volumes are modelled:

- end-expiratory lung volume V_L (FRC)
- series dead space V_{ds}

Further features of the lung model include:

- the lung volume can be washed out (single compartment, mixing ideally)
- CO_2 production with the characteristic 3 phases of the CO_2 trace
- distensibility of the lung
- expiratory peak flow (typical flow pattern of ventilated patients).

Fig. 3.6 shows the principle set up of the lung model. V_L can be adjusted by the addition or removal of water. The series dead space can be changed when desired by

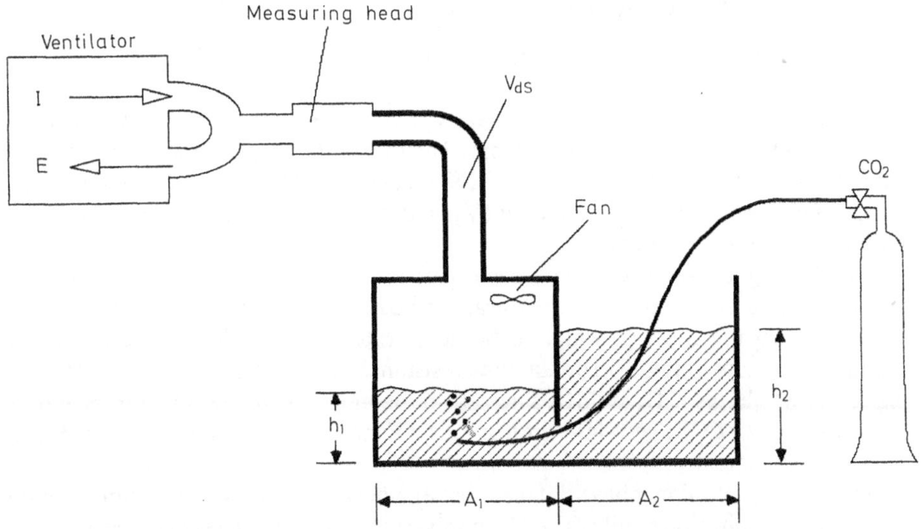

Fig. 3.6. Set up of the lung model. The two chambers of the model with the areas A_1 and A_2 are filled with water (shaded area) and can communicate with each other. The ventilator ventilates the lung volume V_L via the measuring head and serial dead space (V_{ds}). The compression of the gas is only a few cmH_2O (h_2-h_1) and because of the cooling effect of the water, can be assumed to be isothermic. CO_2 production is simulated by introducing pure CO_2. A built-in fan guarantees homogenous mixing

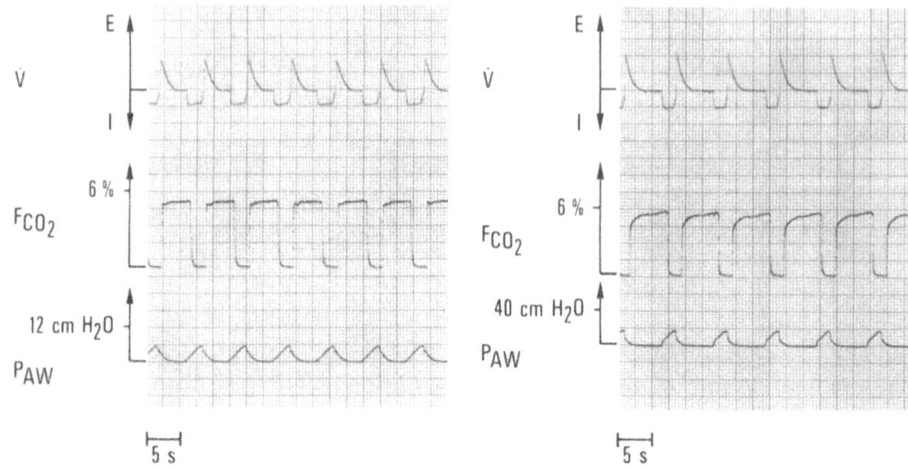

Fig. 3.7. Time course of flow (V′), CO_2-concentration, and airway pressure measured on the model (left curves) and in the patient (right curves) with a V_T of approx. 1000 ml. The end-expiratory pressure of almost 6 cmH_2O confirms the compliance of 168 ml/cmH_2O (6 $cmH_2O \cdot$ 168 ml/cmH_2O = 1008 ml) computed from the geometric measurements

interposing various sized tubes. A built-in ventilator guarantees ideal mixing in the lung volume. The volume distensibility $\Delta V/\Delta P$ of the model lung is given by the two areas A_1 and A_2 in the two chambers according to eq. 3.7.

$$\Delta V/\Delta P = A_1 \cdot A_2/(A_1 + A_2) \cdot k \qquad [3.7]$$

with A_1 and A_2 in cm^2 and $k = 1$ cm/cmH_2O.

The model is a Lucite model (Plexiglas) with base areas $A_1 = 200 \times 150$ mm and $A_2 = 200 \times 190$ mm and a height of 230 mm. The maximum lung volume V_L is 5.6 liters. The volume distensibility is constant at 168 ml/cmH_2O. The flow, CO_2, and pressure signals measured on the Lucite model (V_T = 1000 ml) are shown in Fig. 3.7 together with the signals of a patient.

The flow signal possesses the characteristic feature of the signals measured on the ventilated patient, namely, the expiratory peak flow. In the Lucite model it is dependent on the mass of water to be accelerated, on the flow resistance of the communication opening, and on the flow resistance of the series dead space. The CO_2 signal shows phases I, II, and III. Phase II in the model is steeper than in the patient and corresponds to the response time of the CO_2 measuring instrument (80 ms for the mass spectrometer).

The slightly sloping phase III can be explained as follows: the lung volume decreases in expiration, while the CO_2 supply can be assumed to remain constant or to increase slightly. This causes the CO_2-concentration in V_L to increase, thereby producing the rising "alveolar plateau".

The "airway pressure" rises proportionally to the single breath volume. This linear relationship could be slightly altered by making the area A_2 dependent on the height of the water level h_2. Therefore, even a sigmoid pressure-volume curve (as measured in humans) could be simulated.

This lung model is perfectly suited to simulate oxygen consumption and carbon dioxide production according to the method described by Westenskow et al. (1984). In their method, they introduced CO_2, but diluted the "alveolar" gas with pure nitrogen in order to simulate oxygen uptake. Fig. 3.6 explains how this can be done in the Lucite model. Caution must be taken when analyzing the data, since the added gases cause V_E to become greater than V_I.

3.3.1b Accuracy of Dead Space Determination

To verify the dead space measurement from the CO_2 diagram using the method described in 2.3.5, the Lucite model was wired-up with 7 series dead spaces of different sizes and "ventilated" with 1140 ml breath volume. Twenty breaths were analyzed per situation, while the end-tidal CO_2-concentration remained constant in each situation at 6.7%.

Results: The scatter (± SD) of PIE_{CO_2} was between ± 1.3 ml and ± 2.6 ml. The mean values of PIE_{CO_2} are plotted in Fig. 3.8 against the V_{ds} measured with water. On the average, PIE_{CO_2} was greater than V_{ds}. A regression analysis between V_{ds} and PIE_{CO_2} yielded the relationship $PIE_{CO_2} = 0.993 \cdot V_{ds} + 6.2$ ml ($r^2 = 0.9997$).

3.3.1c Accuracy of the APV Measurement

To determine the accuracy of the APV-measurement, the Lucite model was washed out 6 times with 21 vol% O_2/79 vol% Ar. The tidal volume delivered by the respirator was 1050 ml, the end-tidal CO_2-concentration was between 4.5 and 5.8 vol%. To calculate the lung volume from the amount of the nitrogen measured at the mouth, the nitrogen concentration in the lung at the beginning of the washout must be known. Although the Lucite model was ventilated with 78 vol% N_2, the "intrapulmonary" N_2-concentration was 3.5 to 4.5 vol% lower, since the amount of CO_2 flowing into the

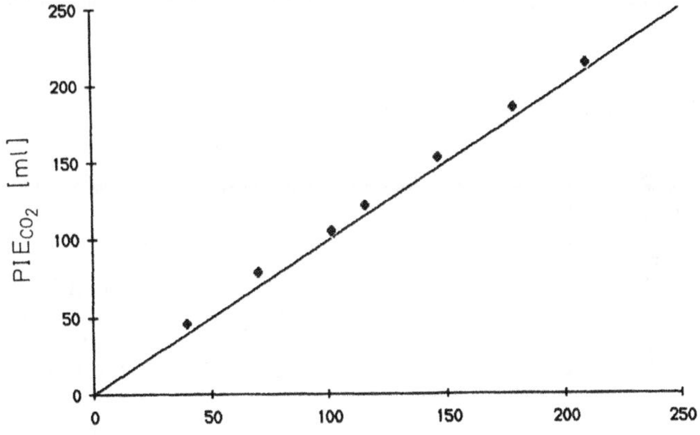

Fig. 3.8. Accuracy of the V_{ds} determination from the CO_2 diagram (PIE_{CO_2}) in the ventilated model. The volume of the tubes which simulate the V_{ds} in the model is plotted against PIE_{CO_2}. The solid line is the identity line

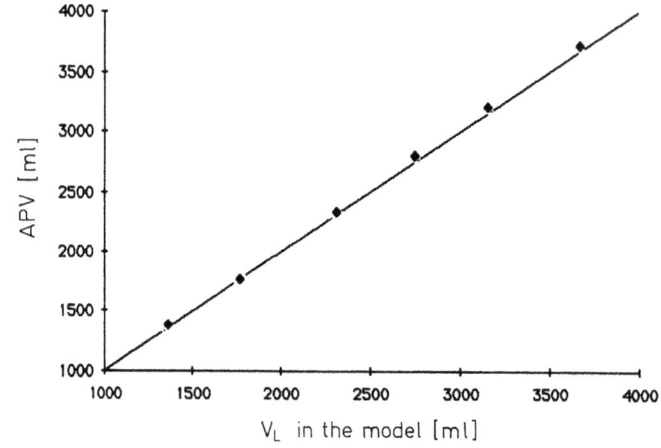

Fig. 3.9. Accuracy of the APV measurement. The lung volume in the model (V_L) is plotted against the APV determined with the nitrogen washout. The solid line is the identity line

lung dilutes the nitrogen. The end-tidal N_2-concentration in the last breath before the washout was taken as a measure of the intrapulmonary N_2-concentration. The raw data were analyzed until the N_2-concentration fell to below 1 vol%.

Results: The "lung volume" was overestimated by an average of 0.62% (Fig. 3.9). The regression analysis between V_L and APV yielded the relationship
$APV = 1.024 \cdot V_L - 40.5$ ml ($r^2 = 0.9999$).

3.3.1d Decay Curve of the Model Lung

Decay curves were calculated for all of the 6 washout maneuvers carried out on the model. One example is illustrated in Fig. 3.10. The decay curve of the model is very close to the calculated, ideal decay curve. The washout efficiencies (WO.eff.) were accordingly high and greater than 98% in every case. The slight deviation from 100% can be explained by the fact that the delay time can only be corrected in steps of 1/60 seconds (sampling rate 60 Hz). This subsequently has an effect on the measurement of PIE_{N_2} and V_{N_2tot}, which in turn affects the decay curve (see Brunner 1985c).

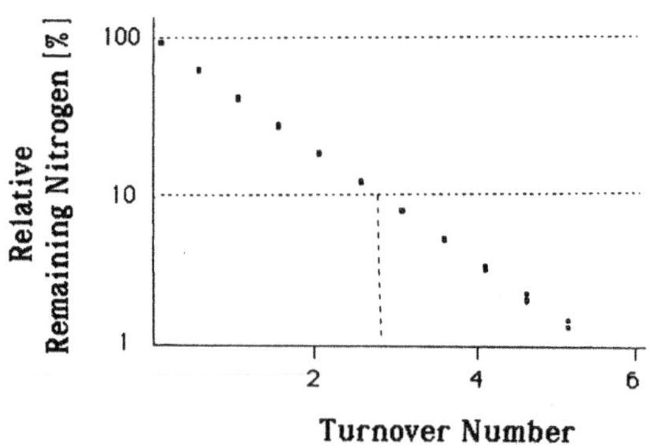

Fig. 3.10. Decay curve of the model lung. The relative remaining nitrogen in the lung (RRN) is logarithmically plotted against the turn over number (TO). The mixing system is ideal (built-in fan) Ideal and experimental curves therefore reached the 10% line at the same TO number

Table 3.2. Scatter of single breath indices in "steady state" (n = 15 times 20)

	V_T ml	PIE_{CO_2} ml	$slope_{CO_2}$ vol%/l	alv.eff.-CO_2 %	C_L ml/cmH_2O	C_{tot} ml/cmH_2O	R_{AW} cmH_2O/l·s
mean value	964	167	0.91	85.3	106	65	5.3
	± 120	± 40	± 0.25	± 2.6	± 30	± 9	± 1.1
mean scatter	19	9	0.14	0.6	2	1	0.2
mean CV	1.8%	5.3%	15.4 %	0.7%	2.1%	2.1%	3.8%

3.3.2. Scatter of the Indices on the Stable Patient in "Steady State"

The scatter of the breath-by-breath measurement of PIE_{CO_2}, $slope_{CO_2}$, alv. eff.-CO_2, C_L, C_{tot}, and R_{AW} was measured on six patients following open heart surgery at various PEEP stages in steady state. Steady state was assumed as soon as F_{CO_2et} remained constant within ± 0.1 vol% for more than 10 minutes.

All of the patients were measured within one hour at PEEP 0 and PEEP 5 cmH_2O. Three patients were additionally measured in the following hour at PEEP 10 cmH_2O. In total, 15 measurements of 20 breaths each were analyzed.

Mean value and standard deviation were determined per group of 20 breaths and are given in the first two lines of Table 3.2. The mean standard deviation was calculated as the square root taken from the mean variance (variance = SD^2). The coefficient of variation (CV) was calculated individually for each measurement and then averaged (mean CV).

The end-tidal CO_2-concentration was constant, its scatter too small to be measured. Just as stable was the breath-by-breath estimation of the efficiency of the CO_2 elimination (alv.eff.-CO_2). The variation coefficient for all of the parameters with the exception of PIE_{CO_2} and $slope_{CO_2}$ was below 4.0%. The largest scatter found for PIE_{CO_2} was ± 15 ml in a group of 20 breaths. This index is very sensitive to spontaneous respiratory activity. If the patient is not sedated deeply enough and is even just a bit excited, the scatter of PIE_{CO_2} increases.

The slope of the alveolar plateau ($slope_{CO_2}$) varied by a maximum of 0.3%/l in a series of 20 breaths. As the $slope_{CO_2}$ in itself is very small, even the slightest change has a substantial effect on the coefficient of variation (CV of 15.4%). The values in Table 3.2 are typical, any deviations occuring in later studies are mentioned in the text.

3.3.3. Reproducibility of the Nitrogen Washout

Practical reasons do not allow a series of several washout maneuvers to be performed in one single patient. One washout maneuver takes 5 to 10 minutes, after which a wash-in has to follow, which also lasts 10 minutes, and one has to wait until this has been completed before the next washout maneuver can begin, and so on. A series of 10 to 20 consecutive washout maneuvers in one patient would take several hours. During this time, the condition of the lungs can change so dramatically that it is impossible to state with certainty whether the washout is reproducible. For this

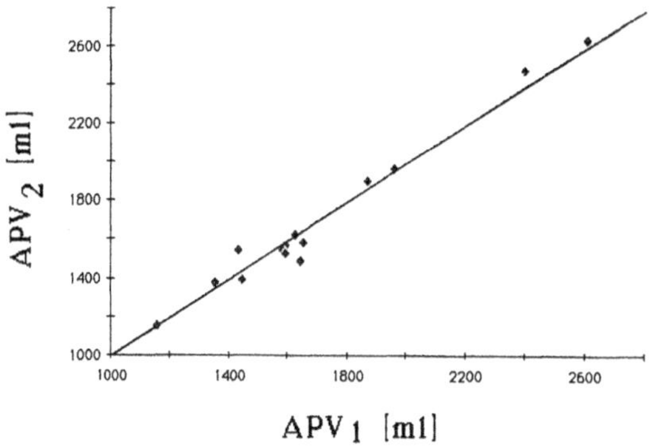

Fig. 3.11. Double measurement of APV. The first measurement (APV_1) is plotted on the abscissa, the second (APV_2) on the ordinate. The identity line is drawn in

reason, we carried out two washouts in each patient within the smallest amount of time possible (so-called double measurement). A total of 18 double measurements (= 36 washout maneuvers) were performed in nine patients following open heart surgery. Four double measurements were disregarded because for clinical reasons the ventilator setting had to be changed by more than 10% between measurements.

The results of the remaining 14 double measurements are given in Fig. 3.11. APV changed by an average of − 14.6 ml ± 66 ml between the first and second measurement. The deviation was a maximum of 159 ml or 10.7% in one case. The second APV deviated from the first by an average of 3.1%.

The mean washout efficiency was 58.6%. The highest efficiency measured was 78.5%, the lowest 32.7%. With three exceptions, the WO.eff. was higher in the second washout than in the first (Fig. 3.12). On the average, there was an improvement of 2.3 ± 1.3% SEM. Several washouts with argon seem to increase the efficiency, however, it is difficult to interpret this on the basis of the data available. It is possible that, contrary to what is generally assumed, changing from nitrogen to argon does in fact disturb the system, as it affects gas mixing and gas distribution in the lungs. It could also be that the two nitrogen washouts took place too soon. If an area is washed out over a period of one hour, it still contains argon at the beginning of the

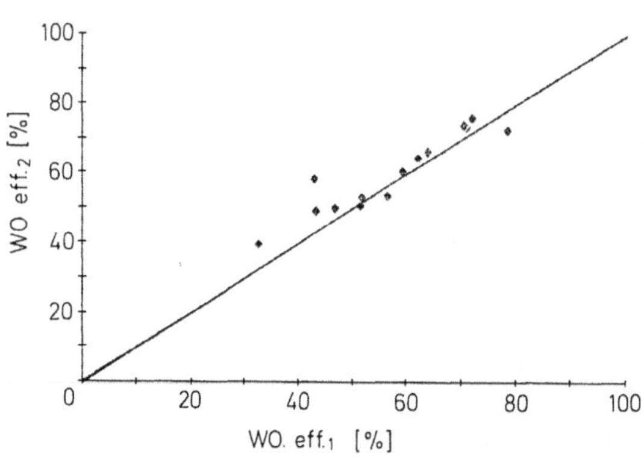

Fig. 3.12. Double measurement of the WO.eff. The first measurement (WO.eff. 1) is plotted on the abscissa, the second (WO.eff. 2) on the ordinate. The identity line is drawn in

second washout, in other words, it is washed-out faster. The APV itself would then necessarily have to become smaller. The mean decrease of − 14.6 ml (0.8%), however, is too small to confirm this hypothesis.

Nota Bene:

- The absolute accuracy of the measuring system is limited by the accuracy of the pneumotachograph.
- To test the entire measuring system, including the software, an adequate physical model that can be ventilated is essential.
- The accuracy of the measuring values and indices can be tested on the physical model. The precision of the indices must be measured on the patient.
- The accuracy of all of the indices is at least 5%, and in most measurements even better than 3%. The precision falls in the same range.

3.4. Use of the Measuring System in the Intensive Care Unit

If the measuring system is to be used in clinical practice, the conditions discussed in Chapter 1 have to be fulfilled. The diagram in Fig. 3.13 illustrates our set up for

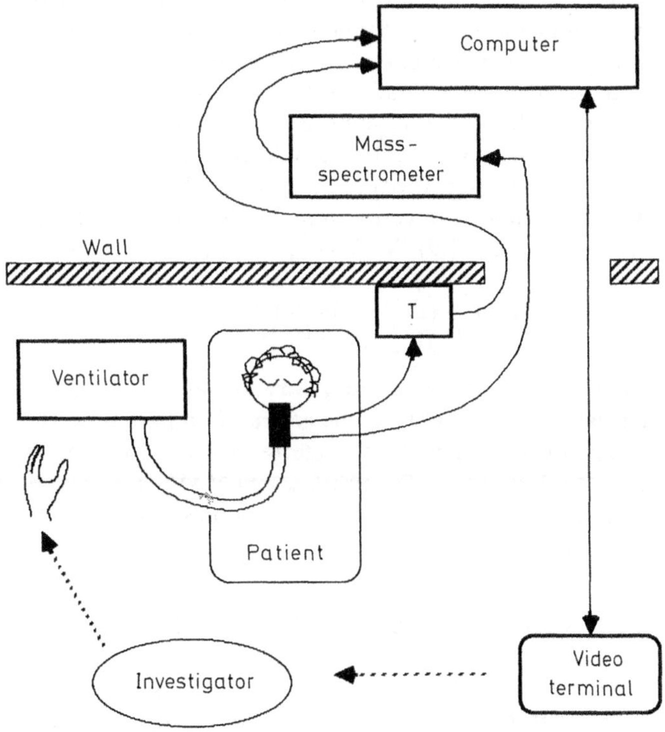

Fig. 3.13. Set up for measurement in intensive care patient. The arrows show in which direction the information flows and suggest a possible control system. The dashed lines are a reminder that it is not a closed loop system. The information is conveyed by the investigator, or by the doctor or nurse

measurements in the intensive care patient. Only the measuring head and the transducers (T) are located near the patient. The mass spectrometer and the computer are located in the adjacent room. All of the electric lines as well as the capillary tube to the mass spectrometer are led through a hole in the wall. The number of devices around the bed is minimal, there is no interference with patient care. The investigator immediately receives the evaluation of the measurement on the computer terminal. In this way, he or she can check whether the protocol is being adhered to and can adjust the ventilator accordingly. By arranging the room in the shape of a star, measurements could be carried out alternatively at three stationary beds without moving the mass spectrometer or the computer.

3.5. Sensitivity of the Indices in the Presence of Acute Pathological Changes in the Lungs: Case Studies

All of the patients described in this chapter underwent open heart surgery with extracorporeal circulation and were treated postoperatively in the intensive care unit. Postoperatively they were paralyzed with pancuronium and sedated with morphine and diazepam. Some measurements were performed in steady state, which was assumed as soon as the end-tidal concentration of CO_2 remained constant at $\pm$ 0.1 vol% for at least 10 minutes. Once set, the ventilator settings remained unchanged throughout the investigation with the exception of T_E, which was adjusted to keep F_{CO_2}-et constant.

Three case studies are presented in the following. In the first two cases, two sets of measurements were performed. A set of measurements included the analysis of the lung function indices described in Chapter 2 using twenty repetitive breaths in steady state, a nitrogen washout, and arterial and mixed venous blood gases. In case 3, the ability to detect breath-by-breath variations is demonstrated. Only one set of measurements was performed to describe the control condition.

To assess the sensitivity of the indices, PIE_{CO_2}, alv.eff. $_{CO_2}$, APV, C_L, C_{tot}, and R_{AW} were compared with conventional indices in cases 1 and 2. Conventional methods refer to C_{eff} (eq. 3.9), V_D/V_T (eq. 3.8), as well as the analysis of arterial blood gas (ABG).

$$V_D/V_T = (P_{aCO_2} - P_{\bar{E}CO_2})/P_{aCO_2} \quad [3.8]$$

$$C_{eff} = V_T/(P_{awoEI} - PEEP) \quad [3.9]$$

where P_{awoEI} is the end-inspiratory airway pressure, $P_{\bar{E}CO_2}$ is the mixed expiratory partial pressure, and P_{aCO_2} the arterial CO_2 partial pressure.

The word "conventional" referred to here does not imply "use of these values in daily clinical practice", since C_{eff} and V_D/V_T are only measured in the exceptional case. By "conventional", is meant that these values were calculated according to text book definitions.

3.5.1. Case 1: Hypovolemia Before and After Correction

Patient F.K.J. (50 years, male, 74 kg, 174 cm) developed severe postoperative bleeding from the operative site as a result of a clotting disturbance with subsequent

Table 3.3. Balance sheet of blood loss and transfusion

time	cumulative volume of blood loss	cumulative volume of transfusion	balance of blood loss and transfusion
4:55	150 ml	180 ml	+ 30 ml
6:40	670 ml	2010 ml	+ 1340 ml
10:20	1510 ml	2830 ml	+ 1320 ml

hemodynamically significant hypovolemia. At 4:55 PM, the following mean blood pressures were noted: aorta (AoP) = 100 mmHg, right atrium (RAP) = 2 mmHg, left atrium (LAP) = 3 mmHg and pulmonary artery (PAP) = 12 mmHg. The first set of measurement was performed under these hypovolemic conditions (Table 3.3).

The clotting disturbance was then remedied and the hypovolemia was treated with a blood infusion. After the positive fluid balance was established, the clinical signs of hypovolemia disappeared and circulation became stable.

At 6:40 PM the mean blood pressures were as follows: MBP = 101, RAP = 5, LAP = 4, and PAP = 15 mmHg. The second set of measurements was done during this hemodynamically stable period.

3.5.1a Results

The ventilator settings in fact remained constant at V_T = 1200 ml, V'_I = 410 ml/sec, EIP = 0, and PEEP = 0. The ABG changed only slightly. P_{aCO_2} dropped from 43.5 to 37.5 mmHg and P_{aO_2} rose from 167 to 176 mmHg. The other results (indices) are compiled in Table 3.4.

FRC and consequently APV are increased during hypovolemia (Rehder 1975). Gray (1978) supported evidence that this is brought about by decreased pulmonary vascular blood volume. In his study on dogs, he observed a 10% increase during hypervolemia caused by inflating a balloon in the left atrium, which in turn brought about an increase in pulmonary vascular blood volume. In addition to the decrease in FRC, this maneuver also decreased the lung compliance, which was also observed by Hauge (1975) in rabbits. Our results show the same. It is very likely that due to the treatment of hypovolemia, the pulmonary vascular volume was increased, which

Table 3.4. Results during hypovolemia (1st measurement) and after successful therapy (2nd measurement). Conventional indices are in bold print

	PIE_{CO_2} ml	alv.eff.-CO_2 %	**V_D/V_T**	APV ml	C_L ml/cmH_2O	C_{tot} ml/cmH_2O	R_{AW} cmH_2O/l·s	**C_{eff} ml/cmH_2O**
1. Measurement	164	79.2	**0.31**	2728	176	91	5.4	**80**
2. Measurement	184	70.9	**0.40**	2106	144	92	5.7	**84**
Difference	20	− 8.3	**0.09**	− 622	− 32	1	0.3	**4**

caused an increase in lung stiffness (Basch 1887). This is another possible reason for the marked reduction of the lung volume.

3.5.1b Comparison with Conventional Methods

The results of the conventional analysis are given in Table 3.4 (bold type). C_{eff} remained nearly constant, but in fact the lung compliance decreased by 18%. Therefore, C_{eff} is not a good estimate of lung compliance. V_D/V_T, on the other hand, increased by 27%, which indicates increased dead space ventilation. The V_D/V_T, however, does not provide precise information unless it is discussed with the aid of PIE_{CO_2} (see Chapter 2.4.4). PIE_{CO_2} rises after therapy by 12%, thereby effecting an increase in V_D/V_T. The remaining change in V_D/V_T, expressed as the change in alv.eff. CO_2, is a functional deterioration which only amounts to half the dead space (Table 3.4).

3.5.2. Case 2: Acute Bronchial Constriction Before and After Therapy

G. H. L.'s (66 years, male, 87 kg, 172 cm) condition deteriorated in the first 5 hours after surgery. He had no past history of any lung disease. The arterial oxygen partial pressure fell to 59.3 torr and auscultation revealed inspiratory and expiratory wheezes. No secretion could be aspirated by tracheo-bronchial suctioning, even after saline lavage. The first set of measurements was taken in this state.

Since acute bronchial constriction was assumed to be accompanied by a bronchial mucosal edema, the patient was treated with aminophylline and prednisolone (high dosage, intravenous). Aminophylline has a dilating effect on the bronchi (and on the vessels), which relieves the bronchial spasm. Prednisolone reduces the mucosal edema and the formation of secretions. A second set of measurements was performed after 70 minutes.

3.5.2a Results

The respirator setting ($V_T = 1180$ ml, $V'_I = 450$ ml/sec, EIP = 0, and PEEP = 0) and P_{aCO_2} (36 mmHg) remained unchanged, while the P_{aO_2} increased considerably (from 59 to 72 mmHg). The results (indices) are summarized in Table 3.5.

Table 3.5. Results during acute asthma (1st measurement) and after successful therapy (2nd measurement). Conventional indices are in bold print

	PIE_{CO_2} ml	alv.eff.-CO_2 %	**V_D/V_T**	APV ml	C_L ml/cmH_2O	C_{tot} ml/cmH_2O	R_{AW} cmH_2O/l·s	**C_{eff} ml/cmH_2O**
1. Measurement	83	69.7	**0.35**	1928	87	56	25.7	**37**
2. Measurement	208	79.6	**0.34**	1879	142	75	11.5	**65**
Difference	125	9.9	**– 0.01**	– 49	55	19	– 14.2	**28**

The broncho-dilating effect of aminophylline is seen in the marked reduction of the airway resistance (R_{AW}). The series dead space (PIE_{CO_2}), which rises at the same time, confirms this finding. The reduction of the edema by prednisolone is revealed partly by the reduction of R_{AW} and partly by the greatly increased distensibility of the lung C_L. It appears that the amount of interstitial fluid has decreased so that the airways have become widened. The resulting improvement of the gas exhange is shown by the marked increase in the alv.eff.-$_{CO_2}$ and the decrease of Q'_S/Q'_T from 29% to 20%.

The CO_2 production was much higher during acute bronchial constriction than in the normal situation, even though the patient was totally paralyzed and deeply sedated. V'_{CO_2} was 127 ml per minute and square meter body surface, as compared with the standard 100 ml/min/m^2 (see application 1). It remains unclear as to why this large amount of CO_2 was produced. It is not likely that the CO_2 stores were emptying, because the end-tidal CO_2-concentration remained constant for several minutes (steady state).

3.5.2b Comparison with Conventional Indices

The results of the conventional analysis are listed in Table 3.5 (bold type). V_D/V_T remained *constant,* which leads one to conclude that there had been no change in the patient's condition. It is possible to make such a misinterpretation, since V_D/V_T consists of components that can show opposite changes: in this example, the series dead space ventilation (PIE_{CO_2}) increased while the alveolar dead space ventilation was reduced (Table 3.5).

The increase in C_{eff} would clinically be considered an improvement, however, this conceals the enormous decrease (55%) of the airway resistance. Thus, C_{eff} does not provide any precise information on the actual changes.

Of the conventional indices, only P_{aO_2} is capable of delivering conclusive evidence. It can be said that the arterial oxygenation has improved in the presence of constant inspiratory O_2-concentration.

3.5.3. Case 3: Transient Processes: Effects of Inflating the Balloon of a Pulmonary Arterial Swan-Ganz Catheter

The conventional indices C_{eff}, V_D/V_T, and ABG are measured "by hand" and are therefore unsuited for both continuous monitoring and for determining rapid, breath-by-breath changes in the function of the lungs. But rapid and transient changes do in fact exist and can be detected. We show this using a model of a small "pulmonary embolism".

3.5.3a A Model for Pulmonary Embolism

Clinically, pulmonary-arterial catheters are used to record the pressure in the pulmonary circulation and to take mixed venous blood samples. Several models (e.g. the modified Swan-Ganz catheter) contain a balloon near the tip, which can be inflated from outside. The actual purpose of inflating the balloon is to record the lung capillary "wedge" pressure to indirectly assess the pressure in the left atrium. By inflating the balloon, a pulmonary-arterial segment is completely blocked. During this time, the perfusion of the respective pulmonary segment by the pulmonary circulation is interrupted. The perfusion is reduced to the bronchial circulation via the systemic circulation, which accounts for less than 1%. The number of affected lung segments depends on the position of the catheter and can be estimated by injecting an x-ray contrast fluid (angiogram in wedged position).

3.5.3b Measurements in the Patient

Patient F. G. E. preoperatively received a pulmonary-arterial Swan-Ganz catheter for hemodynamic monitoring. The position of the catheter tip and the extent of the pulmonary vascular bed can be seen in the chest x-ray in Fig. 3.14 (x-ray after injection of 10 ml contrast medium with balloon inflated). The areas marked in the x-ray (3 to 4 segments) usually receive approximately 17% of the lung perfusion.

The first set of measurements was performed with the balloon deflated. During the second measurement, the balloon was suddenly inflated. Blood samples were taken after approximately 12 breaths and then again after about 27 breaths. The balloon was then deflated again. The same procedure was performed in a second patient, though without blood gas analysis. The results are reproducible.

The wedge maneuver primarily changes the pulmonary perfusion. Nevertheless a chain of consequences results, which with proper methods can be recognized at the mouth.

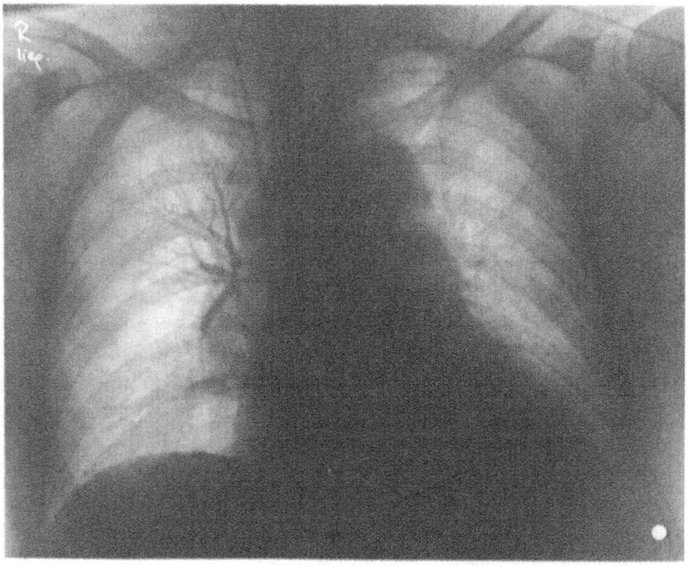

Fig. 3.14. Anterio-posterior chest x-ray of a ventilated patient with Swan-Ganz catheter with balloon inflated (wedge maneuver) in a segment artery of the right lung. The pulmonary-arterial vascular bed blocked by the balloon is shown angiographically. Approximately 17% of the arterial vascular system is blocked

3.5.3c Results

The ventilator settings remained absolutely constant (V_T = 1250 ml, V'_I = 385 ml/sec, EIP = 0, PEEP = 0).

The results before and after balloon inflation are compiled in Table 3.6. All of the indices could be determined in the control condition. Since for clinical reasons the wedge maneuver cannot be carried out over a prolonged period of time, the alv.eff.-CO_2, which can only be measured in steady state, was not measured during the wedge maneuver. Possible changes in the FRC were watched for using V_I and V_E but could not be detected.

The curve of several indices plotted breath-by-breath (Fig. 3.15) documents the transient character of the maneuver.

PEEP and FRC remained constant during the wedge maneuver so that the operating point in the P/V diagram did not change. The reduction of C_L (ΔC_L = 19%) and C_{tot} thus signifies a change in lung properties. This could be caused by increased pulmonary capillary blood volume (Basch 1887, Hauge 1975). Since it is highly unlikely that the cardiac output has changed, the inflated balloon could have effected vascular congestion.

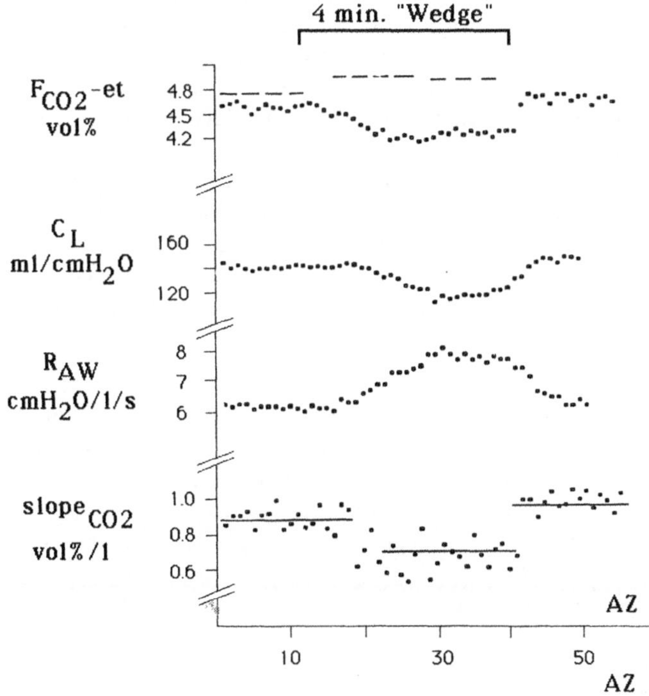

Fig. 3.15. Selected, non-invasively measured indices plotted breath-by-breath. The balloon of the Swan-Ganz catheter is rapidly inflated at breath (SB) 11 and deflated again at SB 40. The maneuver took about 4 minutes. At SB 1, 17, 28, and 38 arterial blood samples were taken (dashed lines in CO_2 curve are the CO_2 partial pressure converted to vol%). Bottom: the solid lines are the mean values of the $slope_{CO_2}$

Table 3.6. Data of patient before balloon inflation (1st measurement) and mean value from 28th to 42nd single breath in Fig. 3.14. CO was 3.7 l/min

	PIE_{CO_2} ml	$slope_{CO_2}$ vol%/l	alv.eff.-CO_2 %	APV ml	C_L ml/cmH_2O	C_{tot} ml/cmH_2O	R_{AW} cmH_2O/l·s
1. Measurement	235	0.90	82.8	1780	139	70	6.1
2. Measurement	236	0.70	*	see Text	116	67	7.6
Difference	1	− 0.20	*	0	− 23	− 3	1.5

* No steady state, alv.eff.-CO_2 cannot be calculated.

The previously perfused and now blocked vascular bed continues to be ventilated and represents the model case of "alveolar dead space ventilation". That is why the expired CO_2-concentration falls (Fig. 3.15). The R_{AW}, on the other hand, increases considerably (+ 25%), which could mean that the change in perfusion effects a redistribution of the ventilation (Severinghaus 1961). The fact that the $F_{CO_2\text{-}et}$ nevertheless dropped, i.e. that the dead space ventilation was not reduced by the redistribution, is difficult to explain. It may be that the duration of the wedge maneuver was too short to have an effect on the $F_{CO_2\text{-}et}$.

A most interesting observation is the time course of the changes. Apparently, a new state is reached after one to two minutes, whereas it is reversed within just a few breaths after the balloon is deflated. These time constants could be attributed to nervous reactions, although hormonal factors cannot be excluded (Jindal 1984 on the dog).

The inflation of the balloon caused a distinct pattern of changes in the indices (Fig. 3.15). The question is whether a beginning pulmonary embolism can be determined in this way. Although this question cannot be answered on the basis of the present study, it can be assumed that similar results are obtained when, instead of a balloon, an embolism blocks the pulmonary vasculature.

Nota Bene:
The clinical examples with acute changes have shown that a differentiated assessment of the gas exchange is not possible using conventional indices. In one example, the assessment even turned out to be wrong: V_D/V_T remained constant in the presence of acute bronchial constriction, while V_{ds} and alv.eff.$_{CO_2}$ were greatly affected. The clinical examples illustrate that with the pulmonary function indices obtained with this measuring system, pulmonary malfunction can be specifically described breath-by-breath, which provides a basis for rational therapy.

4. Application 1: Standard Values During Mechanical Ventilation After Cardiac Surgery

When patients with normal lungs are ventilated, such as for abdominal surgery, the lung volume is already found to be reduced just a few minutes after the anesthesia has been induced. Furthermore, x-ray reveals shadows in the dependent zones, which are interpreted as atelectases (Strandberg 1985). Lung volume reduction has been confirmed by means of N_2 washout, He dilution, and computed tomography. Furthermore, the V'_A/Q' ratios have been found to be changed, the arterial O_2 tension lowered, and the distensibility of the lung reduced. Thus, the term "normal" cannot be applied to values in patients with previously healthy lungs obtained immediately after the beginning of mechanical ventilation, i.e., even before an operation. Values obtained after surgery deserve being called so even less. Nevertheless, patients with previously normal lungs who are mechanically ventilated immediately after a particular operation find themselves in a typical situation. The pulmonary function indices can be considered representative standard values for such a situation. In this chapter we shall report the standard values that were recorded in patients mechanically ventilated after open heart surgery in whom the postoperative course was uneventful.

4.1. Patients and Examination

Our report consists of 28 patients (Table 4.1 in the text and Table 1 in appendix 9), who were included in the study according to the following criteria and who were not excluded postoperatively:

- the patient's history showed no indication of bronchial or pulmonary pathology, nor of transitory or chronic pulmonary edema,
- the preoperative physical examination showed no indication of pulmonary or bronchial pathology, nor of pulmonary-venous congestion or pulmonary-arterial hypertension,
- the preoperative thorax x-ray (posterior-anterior and with the patient standing with the x-ray plate parallel to his left side) showed no signs of pulmonary pathology or of pulmonary congestion,
- in the preoperative hemodynamic examination under stable medication therapy, the left-ventricular end-diastolic filling pressure (LVEDP) – in the case of mitral stenosis, the mean pulmonary capillary wedge pressure – was below 22 mmHg,
- anesthesia, operation, immediate and later postoperative course were uneventful.

The average age of the patients was 58 years (± 8). The group of patients seems to be a random sample survey according to weight (b.w.), height (b.h.), and hence body surface (b.s.),

$$b.s. = \exp[\ln(0.0235) + 0.42246 \cdot \ln(b.h.) + 0.51456 \cdot \ln(b.w.)] \qquad [4.1]$$

(according to Gehan 1970). The sex distribution (22 m, 4 f) is typical for heart surgery patients.

The surgical indications and the operations are described for each case in Table 1 in appendix 9. Twenty-two patients underwent revascularization surgery for chronic coronary heart disease with therapy-resistant angina pectoris (AP) (ACB−n = aortocoronary bypass, n = number of peripheral graft anastomosis). In two of these cases there was also an impending infarction (IMP), in one case, an aneurysm in the left ventricle (ANEU), which was resected (Aneu); there were 10 cases of posttransmural myocardial infarction (pMIF), one case of additional acute mitral insufficiency (MI), which required mitral valve replacement; in one case a periaortal sympathectomy (Symp) had been performed. In six patients the valvular lesion was the major ailment (MS = mitral stenosis; AI = aortic incompetence; AIS = combined aortic stenosis and incompetence); in two cases the aortic valve was replaced (AVR), in two cases the mitral valve (MVR), and in two cases the aortic and the mitral valves (AVR + MVR).

All of the patients received arterial cannulae for arterial blood samples and arterial blood pressure recordings. The right atrium and the left atrium were cannulated for pressure measurements and the pulmonary artery for pressure measurements and blood samples. These four pressures were converted with pressure transducers (Statham P23Db) and measured with pressure-measurement-amplifiers (Hellige, Freiburg/Br., FRG).

All of the patients were extubated no later than 24 hours after the end of the operation and were released from the hospital usually within 12 days, at the most 3 weeks later. The protocol of this investigation was approved by the ethics commission of the Department of Surgery, University of Basel.

The patients were examined as soon as the core temperature (rectal temperature 35.9°C ± 1.5) and the peripheral temperature had nearly returned to normal, and as soon as any primary hypovolemia had been corrected. This was on the average 3.4 (± 1.9) hours after the end of the operation.

The cardiac rhythm (if necessary, under atrial, sequential, or ventricular pacing using temporary electrodes) was regular, the average heart rate (heart-R) was 93 (± 9)/min. Circulation was stable. The mean arterial blood pressure (art. BP) was an average of 85 mmHg (± 13), the mean pressure in the right atrium (RAP) an average of 8 mmHg (± 2), in the left atrium (LAP) an average of 10 mmHg (± 3), and in the pulmonary artery (PAP) an average of 21 mmHg (± 4). Altogether, the values were typical for a controlled situation following heart surgery.

Table 4.1. Table of standard values. The values or their log (*) were normally distributed ($p > 0.1$, skewness $\pm$ 0.5), 2 outliers were tolerated (expected V_{dS}, see Hart 1963)

Index	absolute mean ± SD	normalized mean ± SD
Age	58.2 ± 8.4 y	
Weight (b.w.)	79.6 ± 11.2 kg	
Height (b.h.)	172.3 ± 9.0 cm	
Body Surf. (b.s.)	1.97 ± 0.17 m^2	
LVEDP	14.0 ± 4.7 mmHg	
Time of invest.*	3.4 ± 1.9 h after end of operation	
Body Temp.*	35.9 ± 1.5°C	
Heart Rate	93.0 ± 9.0 beats/min	
mean arterial BP	85.0 ± 13.0 mmHg	
RAP	7.7 ± 2.5 mmHg	
LAP	9.6 ± 3.4 mmHg	
PAP	20.7 ± 3.6 mmHg	
RR	7.3 ± 1.0 breaths/min	
V_T	1078.0 ± 145.0 ml	13.7 ± 2.3 per kg b.w.
V'_I*	389.0 ± 133.0 ml/sec	4.9 ± 1.7 per kg b.w.
I:E	36.0 ± 12.0% of T_I+T_E	
MV	7.8 ± 1.4 l/min	
P_{awo}-max	19.4 ± 3.2 cmH_2O	
P_{awo}-mean	5.4 ± 1.6 cmH_2O	
C_{tot}	64.0 ± 12.0 ml/cmH_2O	37.3 ± 6.5 per m b.h.
C_L*	121.0 ± 41.0 ml/cmH_2O	70.1 ± 23.2 per liter APV
$R_{AW} + R_L$	7.2 ± 2.5 $cmH_2O/l \cdot s$	4.3 ± 2.3 without R_{TUBUS}
R_{tot}	8.2 ± 2.3 $cmH_2O/l \cdot s$	5.4 ± 2.0 without R_{TUBUS}
APV	1825.0 ± 656.0 ml	90.6 ± 30.0% of expected
eff_{WO}	58.5 ± 15.8%	
M0*	0.69 ± 0.13	
M1	1.74 ± 0.25	
M2	5.63 ± 1.59	
V'_{CO_2}	194.0 ± 35.0 ml/min	2.5 ± 0.5 per kg b.w.
V'_{O_2}	228.0 ± 45.0 ml/min	2.9 ± 0.6 per kg b.w.
RQ	0.86 ± 0.07	
F_{CO_2}-et	5.0 ± 0.6 vol%	
V_D-Bohr	307.0 ± 86.0 ml	
V_D-Physiol.*	346.0 ± 99.0 ml	
V_D/V_T	32.0 ± 6.0%	
P_{aCO_2}	36.2 ± 3.6 mmHg	
P_{aO_2}	125.6 ± 33.6 mmHg	
P_{cO_2}	235.6 ± 10.2 mmHg	
pH	7.43 ± 0.05	
Hb	12.5 ± 1.0 g/100 ml	
S_{vO_2}	62.8 ± 7.3%	
V_{dS}	167.0 ± 50.0 ml	115.0 ± 36.0% of expected
SD Phase 2*	59.0 ± 15.0 ml	
$slope_{CO_2}$*	1.12 ± 0.43 vol%/l	
alv.eff.-CO_2	80.7 ± 6.7%	
Q'_S/Q'_T*	8.3 ± 4.8%	
C.O.*	3.6 ± 0.9 l/min	1.8 ± 0.5 per m^2 b.s.
a-vD_{O_2}	6.5 ± 1.2 ml/l	

4.2. The Ventilation

In all of the patients, the muscular paralysis (pancuronium) and the sedation/analgesia (diazepam/morphine) were continued up until immediately before the investigation, so that all of them had muscular paralysis and were deeply sedated. The patients were therefore absolutely passive during ventilation, which was confirmed by the airway and esophageal pressure curves.

All of the patients were ventilated under volume control with the Monaghan M-250 ventilator. The attending physician was given the following guidelines for the primary setting of the ventilator:

RR:	8/min
V:	12 ml/kg
T_I:	20 to 50% of the $T_I + T_E$
V':	5 ml/sec/kg
EIP:	0 sec
PEEP:	0 cmH_2O
F_{IO_2}:	0.4

The RR was varied by shortening or lengthening the T_E in order to keep the P_{aCO_2} between 30 and 40 mmHg. Steady state was assumed when the ventilatory variables $F_{CO_2\text{-}et}$ remained stable for 20 minutes. All of the investigations were carried out under steady state conditions.

The results of this investigation revealed that PEEP and F_{IO_2} had been precisely observed and that the EIP was less than 0.1 sec in every case. The actual values of the ventilatory setting were

RR:	7.3/min ± 1.0
V_T:	13.7 ml/kg ± 2.3
V':	4.9 ml/sec/kg ± 1.7

and hence a minute volume of

MV: 7.8 L/min ± 1.4

4.3. Breathing Mechanics

The maximum airway pressure (P_{awo}-max) was 19.4 cmH_2O ± 3.2. The mean airway pressure (P_{awo}-mean), however, was only 5.4 cmH_2O (± 12) as a consequence of the low RR. The compliance of the total ventilatory system (C_{tot}) was 64 ml/cmH_2O (± 12), that of the lung (C_L) twice as high at 121 ml/cmH_2O (± 41). In neither of the two indices could the relative scatter be reduced by standardization (with respect to height or APV). The total airway resistance (R_{tot}) is the sum of the airway resistance (R_{AW}), lung resistance (R_L), and tissue resistance (R_{ti}) ($R_{tot} = R_{AW} + R_L + R_{ti}$). Nunn (1977, p. 109) estimated the flow resistance caused by the endotracheal tube (R_{tube}) by taking the inner diameter of the tube, leaving the original length unchanged, and assuming a mean flow of 200 ml/sec. Subtracting the tube resistance in each case did not reduce the relative scatter of the mean values of R_{AW}.

In three patients the esophageal pressure was not measured; therefore C_L and R_{AW} + R_L were only determined in 23 patients.

4.4. Accessible Pulmonary Volume

The mean accessible pulmonary volume (APV) was 1825 ml and scattered greatly (± 656). We calculated the expected values for APV as follows: based on the height (b. h. in m), age (A in years), and sex, we determined (according to Quanjer 1977) the expected value of a healthy, spontaneously breathing patient in a sitting position (expected FRC_{sit}):

$$\text{men: expected } FRC_{sit} = 2.34 \cdot \text{b.h.} + 0.009 \cdot A - 1.09 \quad [4.2]$$
$$\text{women: expected } FRC_{sit} = 2.24 \cdot \text{b.h.} + 0.001 \cdot A - 1.0 \quad [4.3]$$

We reduced the expected FRC_{sit} to the FRC expected in the supine position, according to Ibanez 1982 (expected FRC_{sup}):

$$\text{expected } FRC_{sup} = \text{expected } FRC_{sit} \cdot 0.7 \quad [4.4]$$

We further reduced the expected FRC_{sup} to FRC_{vent} for patients under anesthesia ventilated with IPPV by another 15% according to Hewlett 1974.

To our knowledge, using this "tricky" method is the only way to predict the FRC of anesthetized patients in a supine position. Caution: the variation coefficient of this normalization is at least 20%.

The measured mean APV of 1825 ml was 90.6% ± 30 of the expected $FRC_{sup\text{-}vent}$, i.e., the APV, after several hours of anesthesia, after longitudinal splitting of the sternum, and after extensive surgery under hours of extracorporeal circulation, was only an average of 10% lower than the volume that would be assumed to be present immediately after the induction of anesthesia.

A comparison of the APV and the postoperative thorax x-ray taken directly before the examination under unchanged ventilatory conditions (in supine position, anterior-posterior) showed no obvious correlation. On the contrary, some gross contradictions were found: in no case did the thorax x-ray indicate a reduction of lung volume, even in patients with extremely small APV (patients 20 and 21).

4.5. Washout Efficiency and Moment Analysis

The washout curves were analyzed in two different ways: according to the method described in Chapter 2.3.6 (WO.eff.) and following a suggestion made by Gerald Saidel (1975). Saidel's analysis, which is model-free, is carried out in the diagram $F_{N_2\text{-et}}$ vs. dilution number (TO number). Saidel's original method does not take into consideration that not the entire tidal volume but only the tidal volume reduced by the series dead space can contribute to the N_2 washout.

$$Y(k) = F_{N_2\text{-et}}(k)/F_{N_2}^{O} \quad k = 0, 1, 2, \ldots \text{ breaths} \quad [4.5]$$
$$TO(k) = \Sigma V_E(k)/FRC \quad [4.6]$$
$$\mu_r = \Sigma Y(k) \cdot [TO(k+1) - TO(k)] \cdot TO(k)^r$$
$$\text{with } r = 0, 1 \text{ and } 2 \quad [4.7]$$

Since (1) the end-tidal value varies greatly (cardiogenic oscillations) and (2) the TO number is dependent on the volume of each external dead space and on the volume of V_{dS} measured in each breath, an extended standardization possibly achieves better results, i.e., less scatter. The following standardization is suggested (with the disadvantage, however, that the analysis is no longer model-free), with F_{N_2alv} as the mean alveolar N_2 concentration, calculated as $V_{N_2}(k)/[V_E(k)-V_{dS}(k)]$.

$$Y(k) = F_{N_2alv}(k)/F_{N_2}{}^{O} \quad k = 0, 1, 2, \ldots \qquad [4.8]$$
$$TO(k) = \Sigma[V_E(k) - V_{dS}(k) - V_{ext}(k)]/[APV + V_{ext}(k)] \qquad [4.9]$$
$$\mu_r = \Sigma Y(k) \cdot [TO(k+1) - TO(k)] \cdot TO(k)^r$$
$$\text{with } r = 0, 1 \text{ and } 2 \qquad [4.10]$$

The summation takes place from TO = 0 to TO ≥ 8. Only the first 3 moments μ_0, μ_1, and μ_2 were calculated. The standardized moments M0 = μ_0, M1 = μ_1/μ_0, and M2 = μ_2/μ_0 are given in the tables.

Washout efficiency:

Two major differences can be found when our values of the WO.eff. are compared with Cummings and Semple's vent.eff$_{90\%}$: our values are much more scattered (± 15.8%) and secondly, the WO.eff. is much smaller than the vent.eff$_{90\%}$. It is difficult to explain these differences, as the patients are very different. It is nevertheless amazing that the vent.eff$_{90\%}$, which after all contains the ventilation of V_{dS}, is about 30% larger than the WO.eff. which is calculated after the ventilation of V_{dS} has been mathematically eliminated. One would really have to expect the opposite. Our values correlate much better with those of Kox obtained in a recent study (1982), in which the vent.eff$_{90\%}$ is even less than the WO.eff. reported here. If the influence of V_{dS} is eliminated from Kox's results, the difference dwindles to 11% (69.7 ± 7.9%). This difference might be a result of the variations among the patients.

Moment analysis:

The mean standardized first moment M1 roughly coincides with the values obtained in healthy non-smokers. The scatter, on the other hand, resembles that of smokers (Saniie 1979). Caution should be taken when making a comparison with published values. Saidel's original moment analysis contains – as does the vent.eff$_{90\%}$ – the ventilation of V_{dS}, while in our results the measured V_{dS} was mathematically eliminated. This makes a comparison between our results and those published valid only to a certain degree.

4.6. CO_2 Production and O_2 Consumption

We do, in fact, measure CO_2 elimination and O_2 uptake rather than CO_2 production and O_2 consumption. This is done, however, under very strictly observed steady state conditions. We therefore believe that we may regard the measured values as CO_2 production and O_2 consumption. The values are comparable with those measured in a group of 12 probably similar patients (Abdul-Rasool 1983). Interindividual differences in RQ were astonishingly minimal, with 0.86 and a scatter of only ± 0.07 (SD).

4.7. Conventional Indices for CO_2 Exchange

The F_{CO_2} was normal at 5.0 vol% ± 0.6. The V_D/V_T at 32% ± 6 were also normal, but is far smaller than Nunn's estimation (1977, p. 109), according to which the expected V_D/V_T is 52%. As we discussed the shortcomings of the index V_D/V_T in the previous chapter, we do not want to go into it here. For the same reasons, for $V_{D\text{-}BOHR}$ and $V_{D\text{-}physiol}$ we also confine ourselves to our results.

Pulmonary capillary oxygen partial pressure (P_{cO_2}) was calculated according to the equation for ideal alveolar gas (see Summary of Formulas, page 133). Since there was only very little scatter of P_{aCO_2} and barometric pressure, and F_{IO_2} was kept constant, the P_{cO_2} was also normal with only slight scatter.

4.8. Specific indices for CO_2 exchange

The series dead space (V_{dS}) was determined breath-by-breath as a "pre-interface expirate" (PIE), i. e., as the mean value of a distribution (see Chapter 2; Wolff 1984, Brunner 1985). The mean value refers to the first moment of a distribution relative to the zero moment. In a comparable study (Hedenstierna 1975), a V_{dS} of an average of only 115 ml was found in a group of patients with normal lungs shortly after the induction of anesthesia under IPPV with decelerating inspiratory flow and a V_T of 1034 ml ± 130 (method of Fowler 1948). Even if we consider the flow pattern with the particularly low end-inspiratory flow (see Chapter 5), we cannot explain this difference.

Second moment of phase II: the second moment is a measure for the scatter i. e., a measure for the broad phase II, and is calculated analogous to the first moment according to the following formula for the rth moment (i signifies the ith instantaneous value within an expiration, see Fig. 2.12):

$$\mu_r = \Sigma\Delta F(i)/\Delta V(i) \cdot \Delta V(i) \cdot [V(i) + \Delta V(i)/2]^r \qquad r = 0, \text{1st, 2nd moment} \qquad [4.11]$$

$$\mu_2/\mu_0 = 1/\Sigma\Delta F(i) \cdot \Sigma[V(i) + \Delta V(i)/2]^2 \cdot \Delta F(i) \qquad [4.12]$$

$$\text{M2-phase-II} = \sqrt{(\mu_2/\mu_0)} \qquad [4.13]$$

Caution: the second moment of phase II (M2-phase-II) may only be interpreted when the response time of the gas analyzer is sufficiently small. The response time of the mass spectrometer we used was 80 ms and was therefore probably just within the tolerance limits. More exact studies are needed and are absolutely necessary if the M2-phase-II is to be considered a measure for phase II. Furthermore, the impression that phase II represents a Gauss distribution must be avoided. Although here, too, more investigations are needed, a logarithmically normal distribution has to be assumed. The results of the M2-phase-II in Tables 4.1 and the appendix 9 (Table 1) are therefore exclusively for the interested researcher and not for clinical use.

The mean alv.eff. CO_2 is 80.7% ± 6.7. Comparable values cannot be found in the literature.

The mean Q'_S/Q'_T at 8.3% ± 4.8% is slightly higher than the norm. A lower value can hardly be expected under ventilation and after using extracorporeal circulation.

4.9. Cardiac Output

It is striking that the cardiac index, which we determined using the Fick principle has a mean value of only 1.8 l/min/m^2 ($\pm$ 0.5). This value is less than that found by other authors (e.g. Skarvan 1984, Werner 1984) which was measured using the thermodilution method. The difference is so great that it is probably systematic. The fact that a habit and a fact fall together might serve as a plausible explanation:

a) habit: in measuring the cardiac output by the thermodilution method, the blood flow is determined at the measuring site and for the duration of the injection of the cold-signal, and is then projected onto 1 minute; thus, the blood flow is in fact only measured for a period of about 1 second. As the thermodilution measuring method always causes a 15% cardiac output scatter, a mean value from at least 3 measurements is usually given. Anyone can see that in the ventilated patient, the scatter of these 3 values measured during injection in expiration is less than during an injection administered at any random time. Consequently, it has become habit to inject the cold-signal during expiration or even during apnea.
b) fact: when a patient is mechanically ventilated, all of the 4 hemodynamic pressures rise during inspiration. The cardiac output decreases during inspiration and increases again during expiration. This change is more pronounced the more the severe the hypovelemia and/or the cardiac insufficiency, the larger the breath volume, and above all, the lower the ventilatory frequency.

a) and b): the habit described above, to inject the cold-signal during expiration or apnea in order to minimize the scatter of the individual measurements, and the fact described above that under mechanical ventilation the cardiac output increases in expiration, both result in the cardiac output measured by thermodilution under mechanical ventilation being systematically too high. In experimental investigations (with normal cardiac function, normal intravascular volume, normal lung, and normal ventilator setting), the results of the cardiac output determined by thermodilution (in the pulmonary circulation) with the cold-signal injected during inspiration and expiration, respectively, differed by more than 50% (Jansen 1981).

An attempt is often made to get around this immanent weakness associated with measuring the cardiac output under mechanical ventilation by calling the mean value of, let us say, 5 measurements performed during apnea the measured cardiac output value. Most authors then point out that the absolute value of the mean value measured during expiration does not *have* to be correct if only changes between the different mean values measured during expiration are being discussed. This clinically purposeful argument is only valid if the conditions to be compared differ only with respect to the cardiac output and if the relationship between the real mean cardiac output and the cardiac output measured during apnea is constant. But this will not always be the case, and such conditions must not simply be taken for granted, neither when comparing the various ventilation types or settings, nor when comparing the various circulatory conditions.

In eight patients, we measured the cardiac index 14 times by means of both the thermodilution method and our measuring system using the Fick principle with oxygen (Fig. 4.1). The discrepancy is remarkable. The results illustrate the concepts presented above. They warn against trying to calculate the O_2 uptake as a product of

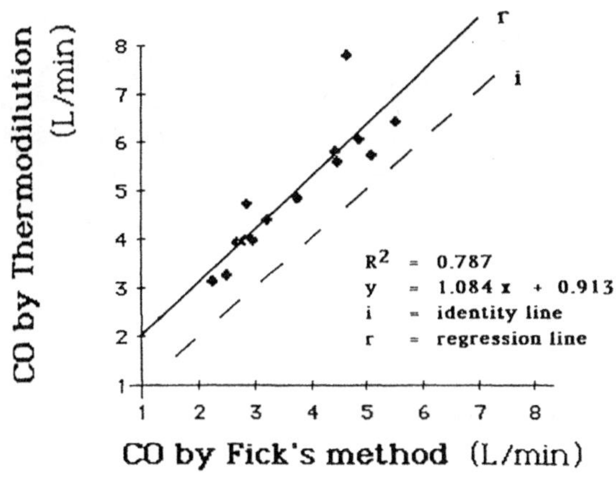

Fig. 4.1. Double measurement of the cardiac output by means of the thermodilution method (Lance Edwards, under injection of the cold signal during apnea) and our measuring system using the Fick principle during ventilation

the cardiac output (determined by thermodilution) and a-vDO_2 (determined with arterial and mixed venous blood gases).

To determine a "real" mean cardiac output during mechanical ventilation by means of the thermodilution method, one would need at least 10 single measurements taken at random times within the breathing cycle. Obviously, under clinical conditions this is out of the question.

4.10. Correlations

There were no correlations between the various indices, with the exception of C_{tot} and C_L. In particular, we found no connection between the APV (in % of the expected FRC) and Q'_S/Q'_T, between the APV and V_{dS}, between alv.eff.-CO_2 and WO.eff. between CO (or CI) and alv.eff.-CO_2, and between LAP or LVEDP and C_L or Q'_S/Q'_T. We conclude that each of these indices carries different information.

Nota Bene:
Under IPPV after an operation that lasts several hours and for which extracorporeal circulation is applied, the pulmonary situation has the following characteristics:

- The pulmonary volume (APV) is reduced by an average of 10% in comparison with the expected value established in other studies for the time immediately after the induction of anesthesia.
- The scatter of APV is large. When the APV is considerably reduced, there is not necessarily a radiological sign of reduced lung volume.
- The mean washout efficiency (WO.eff.) – our measure for the ventilation of the lung – is low (58.5%) and varies considerably (SD = ± 16). The alveolar efficiency for CO_2 elimination (alv.eff$_{CO_2}$) is high (80.7% ± 7). Among the patients examined there is no correlation between the WO.eff. and the alv.eff$_{CO_2}$.

- The cardiac index (CI) measured here with the Fick principle yielded a much lower value (1.73 l/min/m^2 ± 0.45) than those measured by other investigators using thermodilution in comparable situations. We have reason to believe that under mechanical ventilation, methodological problems falsify the measurement carried out with thermodilution.

5. Application II: A Study on Optimizing Mechanical Ventilation

5.1. Problem

Thanks to the progress made in the fields of surgery and anesthesia, older patients and those in poor general condition can now be successfully operated on. During recovery, however, pneumonia is still the major fatal postoperative complication. Advancements in postoperative treatment have reduced the incidence of this serious complication with the result that every year more and more patients are being mechanically ventilated.

During *spontaneous respiration,* the pressure in the lungs is *reduced* by the spontaneous inspiratory movement of the diaphragm and intercostal muscles, which enables the gas to flow into the lungs. During expiration the muscles work in the opposite direction. Under *mechanical ventilation,* on the other hand, the pressure in the lungs is *increased* by the ventilator and the gas is pressed into the lungs. Expiration is passive and is brought about by the relaxation of the lungs and thorax.

In spite of the growing experience with mechanical ventilation and the advancements that have been made, 10–30% of the postoperative and posttraumatic complications are still pulmonary. It is known that the physiological dead space ratio (V_D/V_T) is increased under mechanical ventilation (Campbell 1958) and that the functional residual capacity (FRC) is lowered (Westbrook 1979). Therefore, repeated attempts at improving mechanical ventilation have been made by searching the "optimal" ventilator setting. As will be discussed in the next section, there are discrepancies among the results reported in the literature. The only generally accepted recommendation is ventilation with large tidal volume (V_T) (10–15 ml/kg, Wolff 1983) and low frequency (8–10/min). A closer look reveals that the essential reason for the discrepancies is the differences in the methods used.

5.2. Studies by Other Authors

Several authors have studied the effect of the ventilatory setting on the gas exchange using non-invasive techniques. Some investigated the physiological dead space ventilation (V_D/V_T, eq. 3.8), the effective distensibility (C_{eff}, eq. 3.9), or the functional residual capacity (FRC by means of nitrogen washout). Other authors used invasive methods to measure the influence of the ventilatory setting on the cardiac output CO and on the intrapulmonary right-left shunt (Q'_S/Q'_T). Taking arterial blood is actually invasive, since a catheter has to be inserted. In mechanically ventilated patients, however, an arterial cannula is a routine matter. In this context, therefore, a method is only called "invasive" when *additional* catheters (e.g. pulmonary-arterial) are required.

Some of the results from the literature are summarized in Table 5.1. The example of V_D/V_T when the inspiratory flow V'_I is increased shows that many authors arrived at different results. While Fuleihan (1976) and Connors (1981) were not able to measure any change in V_D/V_T, Baker found an increase (1977b) in V_D/V_T.

Commentary

The authors who studied the I:E ratio did not change only one variable but several at a time, which makes it difficult to interpret their results. The blood gases for the V_D/V_T and Q'_S/Q'_T determination were usually not measured but rather estimated, which required making major assumptions (e.g. constant arteriovenous O_2 difference).

Table 5.1. Influence of ventilator setting on gas exchange: results of other authors measured in normal course with no complications (norm), during severe (ARI) and very severe acute respiratory failure (ARDS), in the dog (D), goat (G), and with lungs damaged by oleic acid (Z-ARDS). Only the results with PEEP < 16 cmH_2O are plotted. – means a negative, + a positive, and 0 no correlation of the index with the ventilator setting (ex. Watson found a decrease of V_D/V_T when I:E was extended)

V_D/V_T	I:E	V_T	V'_I	EIP	PEEP
Nunn 1960		0			
Watson 1962b (norm)	–				
Fairlay 1966 (norm)	–				
Knelson 1970 (D)	–			–	
Sykes 1969 (norm)	?				
Lyager 1968	–			–	
Finlay 1970 (H)	0				
Geiger 1973 (ARI)				–	–
Nosbaum 1974 (ARI)		–			
Hedenstierna 1975b (norm)		–			
Fuleihan 1976 (ARI)			0	–	
Baker 1977b (D)			+		
Suter 1977 (ARI)				–	
Fletcher 1980 (norm)		–			
Connors 1981 (ARI)			0		
Kuckelt 1981 (ARDS)					–
Wolff 1982 (ARDS)					0
Perez-Chada 1983 (G)				–	0
Perez-Chada 1983 (G-ARDS)				–	–
C_{eff}	I:E	V_T	V'_I	EIP	PEEP
Watson 1962a (norm)	+				
Sykes 1969 (norm)	+				
Finlay 1970 (H)	0				
Lyager 1968	0				
Kumar 1970 (ARDS)					0
Falke 1972 (ARDS)					? non-uniform

Table 5.1. (continued)

C_{eff}	I:E	V_T	V'_I	EIP	PEEP
Nosbaum 1974 (ARI)		? non-uniform			
Grimby 1975 (norm)		+			
Baker 1977 (D)			−		
Hobelmann 1977 (ARI)					+
Suter 1978 (ARI)		+			+
Kuckelt 1981 (ARDS)					+
Katz 1981 (ARI)					0
Wolff 1982 (ARDS)					0
Perez-Chade 1983 (G)				0	−
Perez-Chada 1983 (G-ARDS)				+	0

FRC	I:E	V_T	V'_I	EIP	PEEP
Suter 1978 (ARI)					+
Katz 1981 (ARI)					+
Connors 1983 (norm)			−		
Perez-Chada 1983 (G)				0	+
Felton 1984 (norm)			−		

CO	I:E	V_T	V'_I	EIP	PEEP
Sykes 1969 (norm)	0				
Finlay 1970 (H)	0				
Lyager 1968	0				
Kumar 1970 (ARDS)					−
Falke 1972 (ARDS)					0
Geiger 1973 (ARI)				−	−
Baker 1977 (D)			0		
Hobelmann 1977 (ARI)					0
Kuckelt 1981 (ARDS)					0
Wolff 1982 (ARDS)					−
Perez-Chada 1983 (G)				0	0
Perez-Chada 1983 (G-ARDS)				−	−

Q'_S/Q'_T	I:E	V_T	V'_I	EIP	PEEP
Fairlay 1966 (norm)	0				
Sykes 1969 (norm)	0				
Finlay 1970 (H)	−				
Kumar 1970 (ARDS)					−
Cheney 1971 (D)		−		0	−
Geiger 1973 (ARI)				−	−
Nosbaum 1974 (ARI)		? non-uniform			
Baker 1977 (D)			+		
Hobelmann 1977 (ARI)					−
Suter 1977 (ARI)				0	
Connors 1981 (ARI)			0		
Kuckelt 1981 (ARDS)					−
Wolff 1982 (ARDS)					−
Perez-Chada 1983 (G)				0	−
Perez-Chada 1983 (G-ARDS)				0	−
Felton 1984 (norm)			−		

Some authors report on C_{eff}. This compliance was measured by means of various methods, some of which made an additional EIP necessary. Baker (1977a) presents a thorough discussion of the difficulties associated with the various studies. The patients were not always paralyzed, but breathed either synchronously with the machine or were passive. None of the authors made a distinction between these two possibilities and the potential influence on the result of the compliance calculations. Suter (1978) and Katz (1981) measured the FRC only in the initial situation. All of the subsequent FRC changes were calculated from the change in the expiratory volume.

Very few authors measured any additional parameters other than the indices listed in Table 5. Nunn (1960), Hedenstierna (1975b), and Fletcher (1980) assessed the "anatomical" dead space (V_{dAW}) using various methods. Nunn (1960) found V_{dAW} to be unchanged when V_T was increased ($V_T > 350$ ml). According to Hedenstierna (1975b), however, V_{dAW} increased, and Fletcher (1980) still again observed no change. Fletcher found the efficiency of the CO_2 elimination to be dependent on the V_T, whereas in Hedenstierna's study there was no correlation. Hedenstierna, Felton (1984), and Fletcher varied other variables besides V_T and RR. The flow pattern shows a variable pause at the end of expiration. In some cases Fletcher was not able to kep the gas flow constant when the breath volume was increased.

5.3. Hypotheses

The results of these authors show that

1. the conventional methods of analysis have to be refined
2. only *one* variable may be changed per study.

The following hypotheses are tested in the mechanically ventilated patient:

1. The gas exchange is affected by the setting of the ventilator. The changes can be followed on the basis of the indices.
2. There is an optimal ventilatory setting for each and every patient.

Clinical experience with patients with no cardiopulmonary complications (so-called normal course) has shown, however, that it suffices to ventilate patients with an adequate minute volume (MV). The third hypothesis is therefore:

3. In patients with no major cardiopulmonary complications, the breath pattern has no clinically relevant influence on the gas exchange.

5.4. Patients and Methods

5.4.1. Patients

Seventeen patients were examined immediately following open heart surgery. All of the patients (Tables 5.2, 5.4, 5.6, and 5.8) had just previously been operated on using extracorporeal circulation (heart-lung machine, ECC). Postoperative findings revealed no pathological changes in the lungs. The inspiratory O_2-concentration (F_{IO_2}) was 40% in all of the patients studied. The lungs of these patients can therefore be considered "standard" under these special conditions, although no so-called normal physiological situation can be assumed (see also Chapter 4). Such conditions include:

- after ECC
- immediately after heart surgery (aorto-coronary bypass ACB) and/or aortic valve replacement (AVR) and/or mitral valve replacement (MRV)
- slight hypothermia (34–36 °C)
- mechanically ventilated
- sedated and relaxed (with *no* active respiratory movement)
- normovolemia or slight hypovolemia
- normal blood pressures.

In all of the patients, the left ventricular enddiastolic pressure had been preoperatively measured (LVEDP, heart catheter). This pressure exceeded 20 mmHg in a few patients, which represents a permanent change in the lungs. For this reason, these patients are marked with "+" (plus sign).

The applicability and the magnitude of the ventilatory changes were discussed with the attending physician before each investigation. This doctor monitored the effect of the changes on the clinical state of the patient, also during the investigation. For this purpose, the hemodynamics were continuously recorded.

5.4.2. Measurements

The effect of the ventilatory setting on the lung function of the "standard patient" defined above was established in (quasi) steady state. Steady state was assumed when the end-tidal CO_2-concentration remained constant at $\pm$ 0.1 vol% for at least 10 minutes. The first measurement was performed in each patient using the ventilatory setting that had been selected by the medical team in charge (in most cases $V_T = 12–15$ ml per kg body weight). The measured values in this control situation represent the so-called "reference" values of each patient.

Subsequently, one of the 4 ventilatory variables V_T, V'_I, EIP, or PEEP was changed stepwise, while the other 3 were kept constant. V_T ($\pm$ SD) was lowered from 941 ($\pm$ 103) ml to 586 ($\pm$ 119) ml and then raised to 1298 ($\pm$ 131) ml; PEEP was increased from 0 to 6 ($\pm$ 0.7) and then to 11 ($\pm$ 0.8) cmH_2O; V'_I was raised from 386 ($\pm$ 100) ml/sec to 740 ($\pm$ 108) ml/sec and finally an EIP of one second was applied after an EIP of exactly zero. The end-tidal CO_2-concentration was kept constant at $\pm$ 0.2% by adjusting the respiratory rate (by means of T_E). A second measurement was taken within 60 minutes as soon as steady state had set in. In no one patient was the same study done a second time. In the case where 2 or 3 studies were carried out in the same patient, the control measurement was repeated. A total of 40 paired measurements (i.e., 80 single measurements) were performed.

In some patients, for partly technical and partly medical reasons, not all of the measurements and maneuvers could be carried out in full. The gas flow and the CO_2-, N_2-, O_2-, and Ar-concentrations were always measured. Exact data are compiled in the tables of the individual studies.

It is easy to change several ventilatory variables, but it is very difficult to change just one variable and keep the others constant. Since the patient and the ventilator are interrelated, it is necessary to verify each setting on the ventilator on the basis of direct measurements. For this purpose, the measured values were analyzed in real time and the results were displayed on the screen right at the patient's bed during the investigation (Fig. 3.13). Thanks to this feedback, it was possible to keep the V_T constant at 7 ± 30 ml and the V'_I at 5 ± 35 ml/sec (± SD) in the respective studies. The raw data were stored on disk for later analysis.

5.4.3. Presentation of the Results

The results are given in tables, in which the mean values and the standard deviations (SD) of the results in the control situation are listed. The mean values of 10 to 20 single breaths are used for the parameters determined breath-by-breath. Wherever the change in the ventilatory setting produced a statistically significant change in a parameter, this is marked with an asterisk (*).

A change in the parameters was considered statistically significant when

- all of the differences carried the same signs (plus/minus), or
- when the zero hypothesis could be rejected with the Wilcoxon test (fewer than 10 differences) or with the t-test for paired samples (more than 9 differences).

The commonly used significance level of $p < 0.05$ was selected. For presentation in the table, the increase and decrease of V_T were lumped together as a deviation from the reference situation. Similarly, the first PEEP step (6 cmH_2O) was related to ZEEP (0 cmH_2O) and the second PEEP step (11 cmH_2O) to the first (6 cmH_2O) and the results were lumped together. A summary of the most important results is also given in Fig. 5.4. Detailed tables 2.1 to 2.4 are provided in the appendix 9.

5.5. Results and Discussion

5.5.1. Effects of Changing of the Tidal Volume (V_T)

Measurements were performed 30 times in 8 patients (Table 5.2), which yielded 15 paired measurements.

The mean change in V_T was 363 ml or ± 39%. Inspiratory gas flow (318 ml/sec), PEEP (0 cmH_2O), and EIP (< 0.1 sec) remained constant. The results are listed in Table 5.3.

The CO_2-elimination only changed substantially in one patient (+ 65 ml/min or + 30%) and increased on the average only little with increasing V_T. The end-inspiratory pressure rose by 4.4 cmH_2O and the mean airway pressure increased by 10%.

The results show an increased distensibility of the lungs and thorax when the V_T is raised. Similar results were described by Grimby (1975), Nosbaum (1974), and Suter (1978). As PEEP was always zero and APV remained constant, the operating point in the P/V diagram also remained constant. The raised distensibility therefore cannot be explained by a shift of the operating point. A possible explanation is that with increased volume and the therefore necessary pressure, new areas are recruited

Table 5.2. Patients and measurements of the V_T study. + means preoperative LVEDP > 20 mmHg. When Y (N) the data were (not) measured, the N_2 washout was carried out with Ar or 100 vol% O_2

Patient	Age	Sex	Weight	Height	Diag-nosis	P_{awo}	P_{es}	ABG	SAP	N_2 wash-out
N	73	M	76	176	AKB,+	Y	Y	N	N	
A	65	M	82	170	AKB	Y	Y	Y	Y	Ar
Q	61	F	64	144	AKB	Y	Y	N	N	
C	57	F	57	160	MKE,+	N	N	N	N	O_2
R	59	M	68	172	AKE,+	Y	Y	Y	N	
E	58	M	100	175	AKB	Y	Y	Y	N	O_2
F	53	M	71	169	AKB	Y	Y	Y	Y	Ar
S	56	F	61	158	MKE,+	Y	Y	Y	Y	O_2
Mean	59.6		72.4	165.5						
SD	6		13.8	10.8						

Table 5.3. Volume Study. Results in control condition (mean V_T = 1298 ± 131 ml) and paired differences after increase in V_T by 363 ± 123 ml. EIP < 0.1 sec. *: $p < 0.05$

	control mean	± SD	paired differences mean	± SD	
C_{tot} (ml/cmH_2O)	54	10	7.4	3	*
C_L (ml/cmH_2O)	80	24	10	5.1	*
R_{AW} (cmH_2O/l·s)	5.8	2.4	0.4	0.5	*
FRC (ml)	1721	228	40	64	
WO.eff. (%)	75.6	1.9	1.0	4.7	
$P_{awo\text{-}mean}$ (cmH_2O)	5.5	2.0	0.9	0.8	*
$P_{awo\text{-}max}$ (cmH_2O)	19.0	3.5	4.6	1.6	*
P_{awoEI} (cmH_2O)	18.4	3.5	4.8	1.4	*
V_{ds} (ml)	130	21	15	15	*
alv.eff.-CO_2 (%)	86.7	2.8	− 4.5	4.2	*
$F_{CO_2\text{-}et}$ (vol%)	5.1	0.5	0.0	0.2	
V'_{CO_2} (ml/min)	198	41	8	28	
CO (l/min)	3.86	1.2	0.2	0.4	
avD_{O_2} (ml/100 ml)	5.6	2.0	(− 0.2)	0.9	
Q'_S/Q'_T (%)	9.2	4.1	(0.1)	2.8	
P_aO_2 (mmHg)	123	21	12	29	

(Bendixen 1963). This would indeed lead to a larger compliance, but it would also raise the FRC and the APV. APV, however, remained constant here and an opening of the atelectases is therefore an unlikely explanation. Another explanation considers the shape of the P/V curve. Since a larger part of the pressure/volume curve is used when the V_T is increased, the compliance (measured as mean value with our system) can rise due to the upward bend of the P/V curve. APV is so low that the end-expiratory point is probably far below the point of inflection of the P/V curve, making such an effect possible.

Nosbaum (1974), Hedenstierna (1975b), and Fletcher (1980, 1981) found a reduction of V_D/V_T. Our results show non-uniform changes, which supports Nunn and Hill's findings (1960). Since V_D/V_T is the sum of the series and parallel (alveolar) dead spaces, these have to be discussed separately. Fletcher (1980, 1981) did not find any correlation between V_T and series dead space. Our results reveal a definite correlation, which can be explained by the distention of the airways. The series dead space PIE_{CO_2} rose an average of 15 ml (11%). If one traces this increase back to a distention of the airways, the apparent distention of PIE_{CO_2} can be calculated with the aid of P_{awoEI} (4.8 cmH_2O) and amounts to 3.1 ml/cmH_2O. It must be taken into account, however, that the airways are distended by the transmural pressure (P_{awo}-P_{es}) and not by P_{awoEI}. As the transmural pressure does not increase as much as the airway pressure, the distensibility of the airways is higher than that calculated with the aid of P_{awoEI} and PIE.

The reason why Fletcher arrived at a completely different result is probably because he varied not only the V_T but at the same time the EIP. The alv.eff.-CO_2 increased in his study for the same reason, while in ours it decreased with increasing V_T. The reduction of the alv.eff.-CO_2 can be explained by spatial (Rehder 1975) or temporal mismatch (temporal dead space, Cooper 1961) of ventilation and perfusion. Towards the end of inspiration (maximum ventilation), the lung perfusion is reduced as a result of the high alveolar pressure (Jansen 1981, Pinsky 1984). Since the upper (independent) regions are ventilated more than the lower ones, the V'_A/Q' ratio, which is already high in these areas, is increased even more when the V_T is raised. Towards the end of the (passive) expiration, the perfusion and thus the V'_A/Q' increase again. Ventilation and perfusion are thus phase shifted, which means functional dead space ventilation, and can be called temporal dead space ventilation. The larger the tidal (= taking place within a breath) pressure changes, i.e., the larger the V_T, the stronger the effect.

An interesting factor is the increase in the alveolar dead space ventilation, or the reduction of the alv.eff.-CO_2, when the $slope_{CO_2}$ decreases. Fletcher (1980) postulated in contrast that the alv.eff.-CO_2 increases as the alveolar plateau becomes flatter. For illustrative purposes, in Fig. 5.1 *all* of the measurements are plotted in relation to one another. Since the end-tidal CO_2-concentration varied from patient to patient, the $slope_{CO_2}$ was standardized on $F_{CO_2\text{-et}}$. Figure 5.1 clearly shows that there is no correlation between $slope_{CO_2}$ and the alv.eff.-CO_2.

Conclusions

During mechanical ventilation, fresh gas preferably flows into areas with a high V'_A/Q' ratio. The enlargement of the tidal volume reinforces this tendency by impeding the perfusion of independent regions (temporal dead space ventilation). An enlargement of V_T and the resulting pressure increase does not necessarily recruit collapsed areas.

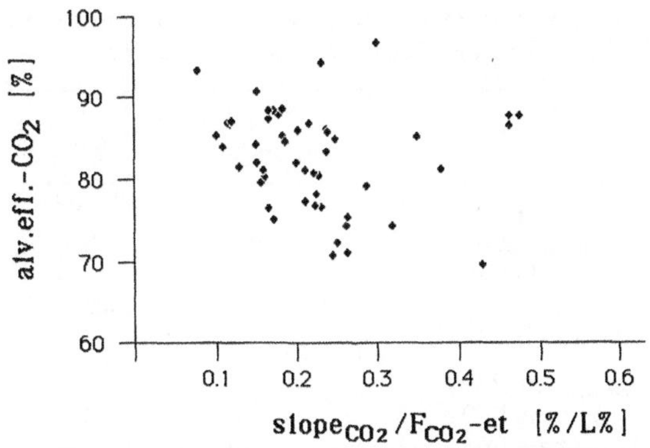

Fig. 5.1. There is no connection between the efficiency of the CO_2 elimination (alv.eff.-CO_2, ordinate) and the standardized increase in the alveolar plateau ($slope_{CO_2}/F_{CO_2\text{-et}}$, abscissa). The results of all of the measurements are plotted, i.e., not only those of the V_T study

Table 5.4. Patients and measurements of the PEEP study. + means preoperative LVEDP > 20 mmHg. When Y (N) the data were (not) measured, the N_2 washout was carried out with Ar or 100 vol% O_2

Patient	Age	Sex	Weight	Height	Diagnosis	P_{awo}	P_{es}	ABG	SAP	N_2 washout
A	73	M	82	170	AKB	Y	Y	Y	Y	Ar
B	61	M	69	164	AKB	Y	Y	Y	Y	Ar
C	57	F	57	160	MKE,+	N	N	N	N	O_2
D	68	M	83	186	MKE,AKB,+	Y	Y	Y	Y	Ar
E	58	M	100	175	AKB	Y	Y	Y	N	Ar
F	53	M	71	169	AKB	Y	Y	Y	Y	Ar
G	60	M	69	173	AKE	Y	Y	Y	Y	Ar
Mean	61.4		75.9	171						
SD	6.9		13.8	8.4						

5.5.2. Effects of the Positive End-expiratory Pressure (PEEP)

In seven patients (Table 5.4), the end-expiratory pressure was increased in a first step from 0 to approximately 6 cmH_2O after the control condition had been measured.

In four of the seven patients, PEEP was raised in a second step another 5 cmH_2O. PEEP-6 was compared with PEEP-0 and PEEP-11 was compared with PEEP-6. This yielded a total of 11 paired measurements. Assuming that the influence of PEEP is independent of whether 0 is increased to 6 cmH_2O or 6 is raised to 11 cmH_2O, all of the paired measurements were lumped together.

The results are compiled in Table 5.5. The inspiratory flow (363 ml/sec), V_T (1002 ml), and the EIP (< 0.1 sec) remained constant. The change in the CO_2 elimination was non-uniform and the mean remained constant. Middle- and end-expiratory airway pressures increased by the same amount as PEEP. Q'_S/Q'_T and CO were reduced only slightly in all of the patients.

Increased ventilatory pressures widens the airways. This is a plausible explanation fo the enlargement of PIE_{CO_2} and also for the reduction of R_{AW}. There is, however, a

Table 5.5. PEEP Study. Results in control condition (ZEEP) and paired differences after increase in PEEP by 5.8 cmH_2O. EIP < 0.1 sec. *: $p < 0.05$

	control mean	± SD	paired differences mean	± SD	
C_{tot} (ml/cmH_2O)	64	9.6	1.4	3.7	
C_L (ml/cmH_2O)	106	34	0.3	15	
R_{AW} ($cmH_2O/l \cdot s$)	5.9	1.4	− 0.6	0.7	*
FRC (ml)	1752	464	671	174	*
WO.eff. (%)	73.6	7.1	− 2.9	9.2	
$P_{awo\text{-}mean}$ (cmH_2O)	4.3	1.0	5.9	0.9	*
$P_{awo\text{-}max}$ (cmH_2O)	16.9	2.3	6.0	1.0	*
P_{awoEI} (cmH_2O)	16.1	2.3	6.2	1.0	*
V_{dS} (ml)	145	39	18	14	*
alv.eff.-CO_2 (%)	85	3.1	0.8	2.5	
$F_{CO_2\text{-}et}$ (vol%)	5.2	0.6	0.1	0.3	
V'_{CO_2} (ml/min)	201	28	1.7	19	
CO (l/min)	4.54	1.0	− 0.3	0.4	*
avD_{O_2} (ml/100 ml)	5.6	1.6	0.2	0.7	
Q'_S/Q'_T (%)	10	4.9	− 1.9	1.0	*
P_aO_2 (mmHg)	120	35	9	8	*

temporal difference between PIE and R_{AW}: while R_{AW} acts during the entire breath, the amount of PIE is determined by the end-inspiratory conditions (see EIP-study below).

In healthy, spontaneously breathing subjects, the airway distensibility is about 3 ml/cmH_2O (Shepard 1957). Hedenstierna (1975 a) studied the dependence of V_{dS} on the end-inspiratory pressure difference P_{awo}-P_{es} in anesthetized and ventilated patients. He varied the pressure difference with the aid of PEEP and found an increase in V_{dS} of 7.5 ml/cmH_2O. In this study, the PIE_{CO_2} rose by an average of 18 ml (13%) and the end-inspiratory airway pressure increased at the same time by 6 cmH_2O. If one carries out the same calculation as in Section 5.5.1 ($\Delta PIE/\Delta P_{awoEI}$), it turns out that PIE_{CO_2} has increased by 3.1 ml/cmH_2O, which confirms the result of the V_T study. And since the pressure surrounding the airways is probably also increased, the distensibility in ventilated patients is also greater than 3.1 ml/cmH_2O.

Shepard (1957) and Martin (1979) described an increase in V_{dS} of 24 ml per liter end-inspiratory lung volume ($V_{L\text{-}EI}$). In the present study, APV increased by 671 ml, which according to those authors has to lead to a PIE increase of 17 ml. This figure correlates well with our measurements (18 ml). Nevertheless, in our patients there was only a shaky relationship between PIE_{CO_2} and $V_{L\text{-}EI}$.

The PIE_{CO_2} of all of the measurements in which $V_{L\text{-}EI}$ was changed by means of V_T or PEEP are plotted in Fig. 5.2. Also included is the correlation given by Martin (1979), which he found in spontaneously breathing volunteers. The following conclusions can be drawn from Fig. 5.2:

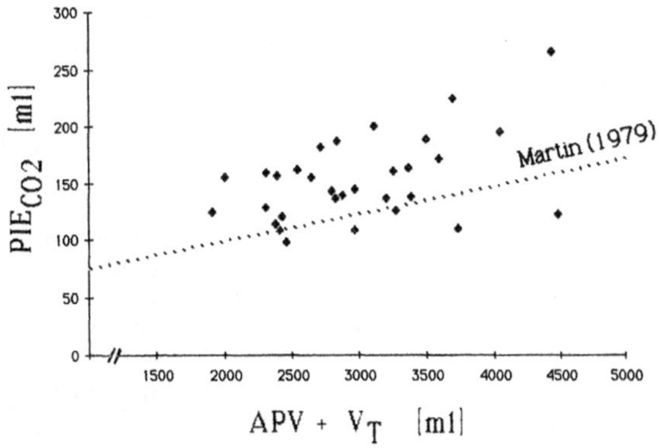

Fig. 5.2. The PIE_{CO_2} of all of measurements in which the $V_{L\text{-}EI}$ was changed by means of V_T and PEEP are plotted on the ordinate against the end-inspiratory lung volume $V_{L\text{-}EI}$ (= APV + V_T). The dotted line shows the regression line found by Martin (1979) in spontaneously breathing volunteers. The regression line found here (*not* drawn) runs parallel to Martin's, but is shifted upwards by 25 ml. The correlation is weak (r = 0.44)

1. The correlation between PIE_{CO_2} and $V_{L\text{-}EI}$ is weak.
2. The quantity PIE_{CO_2} is not primarily determined by $V_{L\text{-}EI}$, although there seems to be a loose correlation. Other factors prevail.

APV increased by 112 ml per cmH_2O PEEP. This increase is much greater than would have been expected on the basis of C_{tot} (64 ml/cmH_2O). Unlike the tidal pressure changes, PEEP remains effective over a longer period of time and thus makes viscoelastic distention possible (Hoppin 1977, Katz 1981). The time constant could be estimated according to observations of inspiratory and expiratory volumes during the measurements: it was in the range of about 1 minute, i.e., after approximately 1 minute, V_I and V_E differed by less than 10%.

PEEP shifts the operating point in the P/V diagram to the right. Since an s-shaped p/V-curve must be assumed, one would expect a concomitant change in compliance. However, we found no significant change in C_{tot} or C_L for the group of patients. In most patients, compliance did not change at all, which would indicate a linear P/V relationship over the PEEP range measured. As was stated above, this is unlikely and therefore we must assume that the P/V relationship has changed due to PEEP. This interpretation questions the conclusions of other authors who based their discussion on the assumption of a fixed P/V curve (Hobelmann 1977, Suter 1978). They found a clear dependence of compliance on PEEP in patients with severe respiratory insufficiency and concluded that PEEP shifts the operating point on a fixed P/V curve. Our results also indicate that the effect of PEEP is more complex in that it does not only alter the operating point, but also the P/V curve itself.

Geiger (1973) and Kuckelt (1981) found a reduction of V_D/V_T in patients with ARI after PEEP increased. The fact that this result cannot be reproduced in our measurements here again shows that a change in the ventilatory setting in a severely pathological situation can have other effects than when this happens during an uncomplicated normal course (Perez-Chada 1983).

Conclusion

PEEP brings about a change in the operating point in the P/V diagram. Further, PEEP effects viscoelastic distension and thus changes the pressure-volume relationship.

Table 5.6. Patients and measurements of the V'_I study. + means preoperative LVEDP > 20 mmHg, S means cardiac insufficiency. When Y (N) the data were (not) measured. The N_2 washout was always carried out with Ar

Patient	Age	Sex	Weight	Height	Diagnosis	P_{awo}	P_{es}	ABG	SAP	N_2 washout
N	60	M	76	176	AKB,+	Y	Y	N	N	
H	65	M	85	172	AKB	Y	Y	Y	Y	Ar
I	50	M	74	174	AKB	Y	Y	Y	Y	Ar
O	61	M	74	169	AKB,+,S	Y	Y	Y	Y	Ar
P	46	F	63	152	AKB,AKE,+	Y	Y	Y	Y	Ar
L	61	M	87	166	AKB	Y	Y	Y	Y	Ar
M	48	M	81	175	AKB	Y	Y	Y	Y	Ar
Mean	55.9		77.2	169.1						
SD	7.6		8.1	8.3						

5.5.3. Effects of the Inspiratory Flow Rate (V'_I)

In 7 patients (Table 5.6), the inspiratory flow was increased by an average of 355 ml/s and an average of 91.1%, respectively (relative to the individual control condition). EIP was always exactly zero and the gas flow remained constant during the entire inspiration (no end-inspiratory decay!). V_T (1132 ml) remained constant and PEEP was 0.

The results (Table 5.7) show that the CO_2 elimination in the presence of high flow dropped slightly. The maximum inspiratory pressure increased, however, the mean airway pressure decreased. The slope for CO_2 rose in every patient.

The marked reduction of the APV (Connors 1983, Felton 1984) can be attributed to the extension of the expiratory time; a low end-expiratory gas flow of, let us say, 10 ml/sec would evade proof, but it would cause the APV to drop by 60 ml in one minute when the T_E was prolonged by 1 sec. Viscoelastic forces which cause low flow could possibly explain the emptying of the lungs during prolonged expiration. This results in a vertical shift of the operating point, i.e., the original P/V curve is left, since the end-expiratory airway pressure remains unchanged. The compliance itself decreases, which was also found by Baker in his experiments on the dog (1977b). The shifting of the operating point alone cannot explain this result. It must be assumed that other mechanical properties of the lungs are made use of when there is high inspiratory flow. The shortened inspiratory time diminishes the viscoelastic distensibility and hence possibly lowers the compliance values.

Increasing the inspiratory flow increases the inspiratory pressures and the airways are thereby distended. When there is a distensibility of 3 to 7 ml/cmH_2O (see above and Hedenstierna 1975a), an increase in P_{awoEI} by 1.5 cmH_2O (Table 5.7) results in an enlargement of PIE by 4 to 10 ml. The results, however, show a PIE_{CO_2} of 29 ml. Therefore, an additional mechanism must be assumed. According to Paiva (1972), in the presence of an increase in flow in the acinus, the SI is shifted towards the end of the acinus and thus PIE is enlarged (see appendix 4, eq. 4.8).

Table 5.7. Flow Study. Results in control condition (mean $V'_I = 386 \pm 100$ ml/sec) and paired differences after increase in V'_I by 355 ± 50 ml/sec. EIP < 0.032 sec. *: $p < 0.05$

	control mean	± SD	paired differences mean	± SD	
C_{tot} (ml/cmH_2O)	69	17	− 1.8	3.2	
C_L (ml/cmH_2O)	112	30	− 7.1	8.6	*
R_{AW} (cmH_2O/l·s)	8	3	0.1	1.3	
FRC (ml)	1805	401	− 235	151	*
WO.eff. (%)	46.6	8.1	1.1	7.0	
$P_{awo\text{-}mean}$ (cmH_2O)	4.8	0.7	− 1.1	0.8	*
$P_{awo\text{-}max}$ (cmH_2O)	19.7	4.0	3.4	0.9	*
P_{awoEI} (cmH_2O)	18.5	3.7	1.5	0.6	*
V_{ds} (ml)	173	28	29	19	*
alv.eff.-CO_2 (%)	80.0	5.3	− 5.0	6.1	*
$F_{CO_2\text{-}et}$ (vol%)	4.9	0.5	0	0.1	
V'_{CO_2} (ml/min)	174	12	− 13	12	*
CO (l/min)	3.54	1.0	− 0.3	0.5	*
avD_{O_2} (ml/100 ml)	6.1	1.4	0.2	0.6	
Q'_S/Q'_T (%)	9.4	7.5	− 1.1	1.3	*
P_aO_2 (mmHg)	124	45	11	15	*

The alv.eff.-CO_2 fell by an average of 5%, i.e., the alveolar dead space ventilation increased. The increase in V'_I thus leads to a redistribution of the inspirate, namely in areas with high V'_A/Q' (dead space). When the flow velocity is high, the gas tends to flow in the direction of small flow-resistance. Since the upper areas have a lower flow resistance than the lower regions (Nunn 1977), and are marked by a high V'_A/Q' (Rehder 1975), it can be concluded that the flow increase causes increased ventilation of the upper areas.

Q'_S/Q'_T both dropped slightly. This finding is contradicted by Baker (1977 on the dog), but it correlates with the Felton's findings (1984), who observed improved oxygenation (increase in P_{aO_2}) with a flow increase. It is a peculiar result that Q'_s/Q'_T is reduced by high inspiratory flow, and cannot be explained with present evidence. The decrease in Q'_s/Q'_T, however, coincides with reduced CO. It is known that a drop of the CO can lower the venous admixture (Q'_S/Q'_T) (Wolff 1972, Steenblock 1976 on the dog, Lemaire 1978). The slight changes in Q'_S/Q'_T might therefore be connected with the reduction of the CO.

Conclusions

High inspiratory flow shifts the stationary interface towards the end of the acinus and thereby enlarges the airway dead space. Independent regions are very likely ventilated when the inspiratory flow is high. The expiratory time, which was prolonged for compensation purposes (for MV = const.) in the presence of high V'_I, allows APV to drop.

Table 5.8. Patients and measurements of EIP study. When Y (N) the data were (not) measured. The N_2 washout was always carried out with Ar

Patient	Age	Sex	Weight	Height	Diag-nosis	P_{awo}	P_{es}	ABG	SAP	N_2 wash-out
H	65	M	85	172	AKB	Y	Y	Y	Y	Ar
I	50	M	74	174	AKB	Y	Y	Y	Y	Ar
K	61	M	89	173	AKB	Y	Y	Y	Y	Ar
L	61	M	87	166	AKB	Y	Y	Y	Y	Ar
M	48	M	81	175	AKB	Y	Y	Y	Y	Ar
F	53	M	71	169	AKB	Y	Y	Y	Y	Ar
Mean	56.3		81.2	171.5						
SD	6.9		7.3	3.4						

5.5.4. Influence of the End-inspiratory Pause (EIP)

Six patients (Table 5.8) were measured without EIP and then with EIP. To intensify the effect of EIP, the inspiratory flow V'_I selected was twice as high as with the other measurements. V'_I (589 ml/sec), V_T (1113 ml), and PEEP (0 cmH_2O) remained constant in this study.

The results are summarized in Table 5.9. End-tidal CO_2-concentration and CO_2 elimination changed only slightly. The maximum inspiratory pressure decreased by

Table 5.9. EIP Study. Results in control condition (EIP < 0.1 sec) and paired differences after having set an EIP of 1 sec. *: $p < 0.05$

	control mean	± SD	paired differences mean	± SD	
C_{tot} (ml/cmH_2O)	71	8.1	2.0	4.1	
C_L (ml/cmH_2O)	119	28	3.2	4.5	
R_{AW} (cmH_2O/l·s)	7.0	1.8	− 0.2	1.2	
FRC (ml)	1753	363	43	233	
WO.eff. (%)	53.3	10.4	− 1.8	6.3	
$P_{awo\text{-}mean}$ (cmH_2O)	3.7	1.1	2.0	0.8	*
$P_{awo\text{-}max}$ (cmH_2O)	20.0	4.6	− 1.4	1.2	*
P_{awoEI} (cmH_2O)	18.0	3.8	− 2.2	1.2	*
V_{ds} (ml)	198	39	− 38	19	*
alv.eff.-CO_2 (%)	79.1	7.0	3.4	1.6	*
$F_{CO_2\text{-}et}$ (vol%)	5.3	0.5	− 0.1	0.1	*
V'_{CO_2} (ml/min)	177	40	12	12	*
CO (l/min)	3.92	1.3	− 0.2	0.8	
avD_{O_2} (ml/100 ml)	5.8	1.6	0.6	2.0	
Q'_S/Q'_T (%)	7.9	5.1	− 1.0	1.1	*
P_aO_2 (mmHg)	138	32	5.0	12.6	

1.4 cmH_2O and the end-inspiratory pressure ("plateau pressure") fell by 2.2 cmH_2O in the presence of EIP. The mean airway pressure, on the other hand, increased by more than 50% (!).

Even though all of the ventilator settings – with the exception of EIP – remained constant, the maximum pressure in the presence of EIP dropped for technical reasons. The ventilator generated a flow pattern which drops slightly during end-inspiration when an EIP is set. The peak pressure, which would have occurred had the inspiratory flow been constant, is not reached. It is assumed that this slight change has no effect on the gas exchange.

When an EIP is made, the expiratory time has to be shortened accordingly in order to keep the alveolar ventilation and the $F_{CO_2\text{-et}}$ constant. The results of the V'_I-study (decreasing APV with prolonged T_E) led to the conclusion that the APV increases with shorter expiratory time. The results in Table 5.9 show, however, that this was not the case in the EIP-study (see Fig. 5.4). This finding concurs with Perez-Chada (1983).

V_D/V_T is greatly reduced by an EIP. The same result has also been published by other authors (Knelson 1970, Lyager 1970, Geiger 1973, Fuleihan 1976, Suter 1977, Perez-Chada 1983). The rôle played by the series and alveolar components, however, has not yet been clarified. Our results show that PIE (when EIP = 1 second) changes relatively more than the alveolar dead spaces. Two main causes of the reduction of PIE (Fowler 1948, Hedenstierna 1975 a) come into question:

a) A pressure gradient from the entrance of the endotracheal tube down to the alveoli builds up due to the inspiratory gas flow. When the expiratory valve opens without EIP, the airways relax and gas is expelled into the atmosphere. If, on the other hand, an EIP is made, the airways relax partially and gas is expelled into the lungs, thereby decreasing the volume of the airways.
b) The increased inspiratory time promotes diffusion. The SI is shifted towards the mouth and diminishes the volume of series dead spaces.

Both mechanisms led to a reduction of PIE_{CO_2} by a mean of 19%.

The reduction of the alveolar dead space ventilation or the increase in the alv.eff.-CO_2 can be explained by the improved match of ventilation and perfusion (West 1969). High inspiratory flow generates intrapulmonary pressure gradients. The EIP allows pressure equilibration, which may lead to a redistribution of the ventilation. Such equilibration processes also have an effect on the alveolar plateau and probably for this reason the $slope_{CO_2}$ is reduced after an EIP.

Table 5.9 shows the results in steady state. As a supplement to this, Fig. 5.3 shows the changes in PIE_{CO_2} and $slope_{CO_2}$ during a breath-by-breath change in EIP. These measurements were carried out in one patient within 8 minutes, during which EIP was altered by hand and at random times between zero and 3 seconds. Afterwards, the length of the EIP in each breath was exactly measured by hand (± 0.05 seconds). Fig. 5.4 shows that PIE_{CO_2} drops the most in the first part of the EIP. The time constant in this first part of the curve is probably less than 0.5 seconds. This result provides evidence that for V_{dS} investigations, the end of inspiration has to be exactly controlled.

The $slope_{CO_2}$, on the other hand, seems to decrease rather linearly with time. The changes between 0 and 1 second are comparable with those between 1 and 2 sec EIP. If the time constants of PIE_{CO_2} and $slope_{CO_2}$ in Fig. 5.3 are compared, it becomes

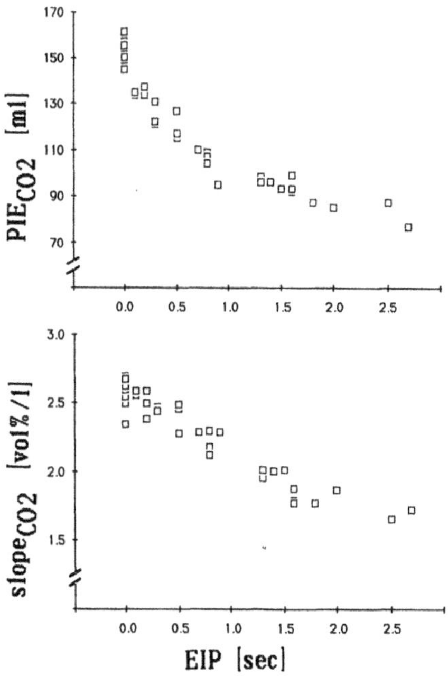

Fig. 5.3. Dependence of V_{ds} (PIE_{CO_2}, left) and the $slope_{CO_2}$ (right) on the length of the EIP measured in a patient. Each point represents the measured value of one breath, the EIP was altered at random. PIE_{CO_2} changes considerably as the EIP becomes shorter (in comparison with $slope_{CO_2}$) and less as the EIP becomes longer

evident that both slow and fast processes take part in the intrapulmonary gas exchange.

Conclusions

An EIP allows a reduction of the dead space ventilation by means of two mechanisms:
1. intrapulmonary pressure equilibration and
2. promotion of diffusion.

5.6. Summarizing Discussion and Conclusions

5.6.1 Non-quantifiable Observations

CO_2 production was constant throughout the studies. During relaxation, metabolic activity is reduced to almost the "basal" level (100 ml/min/m^2 body surface, Nunn 1977) and V'_{CO_2} is thus held constant. The CO_2 production rose as soon as the patient woke up or was no longer sufficiently relaxed. The V'_{CO_2} reacted very sensitively and suggested changes in the patient's condition even where no clinical signs of such changes were evident. The end-tidal CO_2-concentration behaved in a similar manner and proved to be extremely sensitive to changes in the ventilator setting. Furthermore, in one patient, simply a change in the heart rate (with the aid of a pacemaker) caused the $F_{CO_2\text{-et}}$ to drop. It must be assumed that in this case the CO was altered as the heart rate changed.

5.6.2. Optimizing Ventilation

The transport of CO_2 and O_2 is primarily achieved by a certain minute volume, which can be accomplished with a large number of different ventilator settings. The present results, which are summarized in Fig. 5.4, clearly demonstrate that the pulmonary function indices are dependent on ventilatory variables. Most positive influences on the gas exchange, however, were combined with negative influences. With small tidal volumes, for example, the efficiency of the CO_2 elimination improved, but the compliance of the lungs decreased. PEEP improved oxygenation but not the alv.eff.-CO_2. An EIP increased alv.eff.-CO_2, but also increased the mean airway pressure and therefore, along with PEEP, had a potential negative effect on cardiac performance. Although lowering the inspiratory flow promises largely positive effects, oxygenation is improved by an *increase* in the inspiratory flow and not a decrease.

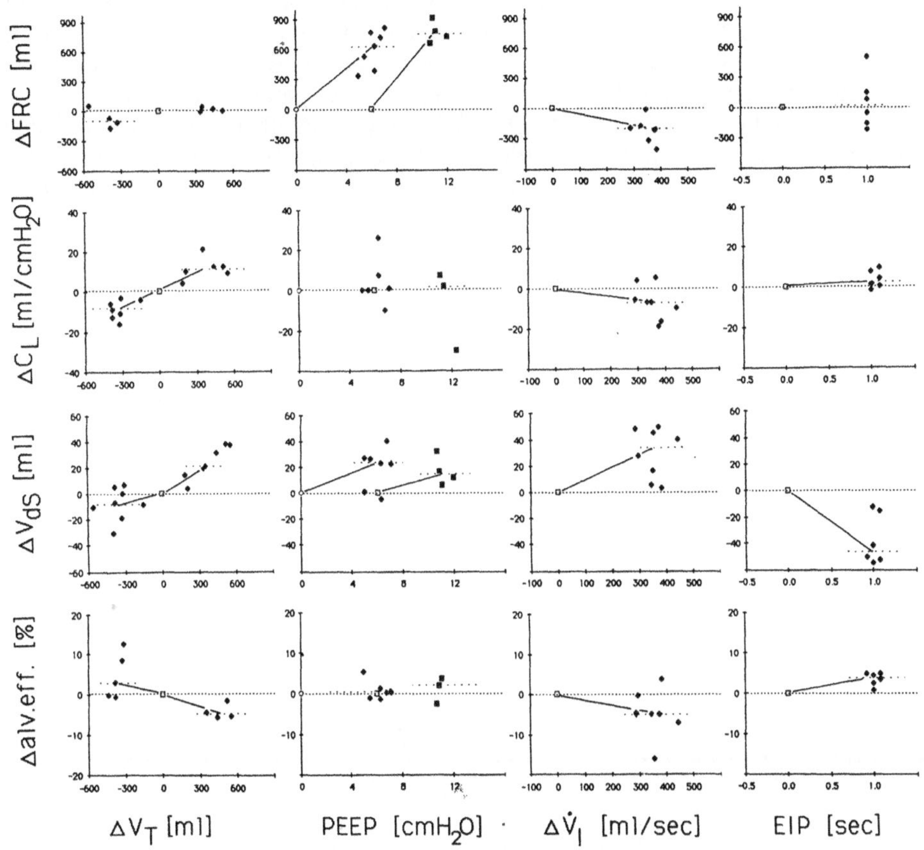

Fig. 5.4. Summary of the most important results. The changes of the ventilatory variables are plotted on the abscissas. Increase and decrease of V_T are represented separately. The PEEP steps are also listed separately. The absolute changes (closed symbols) are plotted on the ordinate relative to the reference situation (open symbols). The median of the paired differences is represented by a dotted line and connected to the reference value by a solid line when the change is statistically significant (see 5.4.3)

These results show that in the patients investigated in our study, there is no ventilator setting that combines all the positive effects. In other words, no optimal pattern of ventilation could be found.

The effect of different ventilator settings can be measured, however, it apparently has no clinical relevance for the gas exchange. A possible explanation for this can be derived using the index for washout efficiency.

5.6.3. Recruitment of Compensatory Mechanisms

The results have demonstrated that during uncomplicated postoperative courses following cardiac surgery, intrapulmonary gas mixing is largely independent of the ventilator setting. No correlation was found between the WO.eff. and the ventilator setting, which quantitatively corroborates Bergman's findings (1969). Each patient had his or her own characteristic WO.eff., which remained relatively constant, but which varied greatly from patient to patient. In 3 patients it was less than 45% and in one patient even dropped to below 30%. The decay curves of this patient were not even close to being straight, as they were in the other patients. That means that a portion of the lungs is washed out only very slowly or, in other words, receives only little ventilation in comparison with its volume (low relative ventilation). This interpretation is confirmed by the N_2 diagram in the late washout of this patient, which shows a pronounced phase IV (Fig. 5.5).

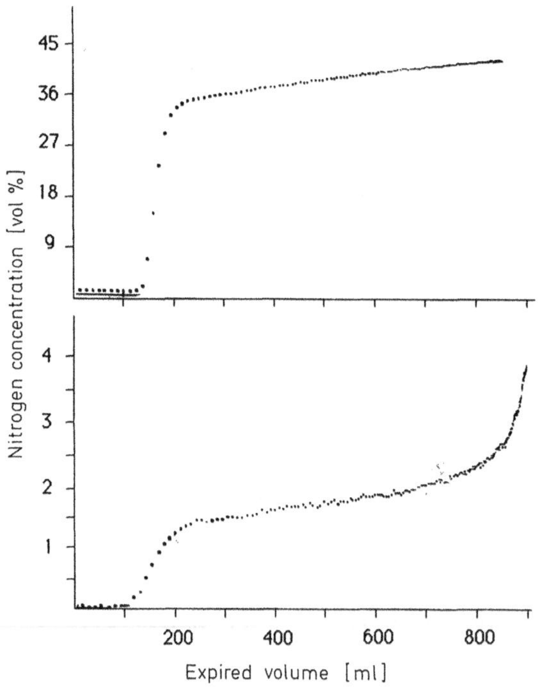

Fig. 5.5. N_2 diagram of the first (above) and eleventh (below) breath in the N_2 washout of a patient. The pronounced phase IV is still concealed at the beginning of the N_2 washout and is not recognizable until the late N_2 washout

Two possible explanations of this marked phase IV can be considered:

1. A leak in the system allows air and hence N_2 to penetrate at the end of expiration. If this were so, CO_2 would have to fall at the end of expiration. This was checked, but it was not observed in any case.
2. A portion of the lungs receives only very low relative ventilation. Furthermore, this portion is ventilated asynchronously, which leads to the formation of phase IV.

In the 3 patients with low WO.eff., the lungs are apparently unevenly ventilated and it would be expected that this affects the CO_2/O_2 gas exchange. In fact, however, neither a high Q'_S/Q'_T nor a reduced alv.eff.-CO_2 could be measured in those patients. Furthermore we found no correlation between WO.eff. and alv.eff.CO_2 (Fig. 5.6). This leads to the conclusion that patients with a normal postoperative course recruit enough compensatory mechanisms so that even under difficult conditions (i.e., in this case a low WO.eff.) an adequate exchange of CO_2 and O_2 is accomplished.

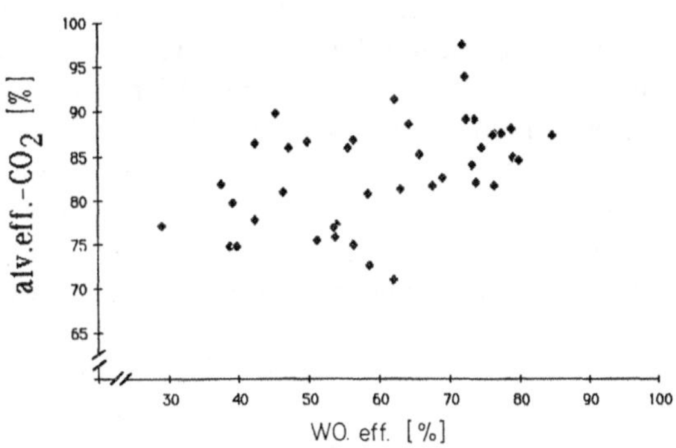

Fig. 5.6. Relationship between the alveolar washout efficiency from the N_2 washout (WO.eff.) and the efficiency of the CO_2 elimination (alv.eff.-CO_2). All of the measured values of all of the patients are plotted. The correlation between WO.eff. and alv.eff. is very weak

Nota Bene:

- Leaving serious false settings out of consideration, the ventilator setting has only a minimal and clinically irrelevant effect on gas exchange in the normal course without complications. This is because the patient is able to recruit enough compensation mechanisms.
- It is likely that for this reason, the full effect of a change in the ventilator setting cannot be measured. At most, a "steady state offset" can be measured as a result of incomplete compensation. The magnitude of such a steady state offset turned out to be negligible in the present investigation. From this we infer that in such patients, there is no rational justification for trying to make subtle ventilator adjustments.
- An optimal setting of the ventilator enables the cardiorespiratory system to adapt to the special conditions that exist under mechanical ventilation.
- On the other hand, it can be presumed that patients with severe respiratory and/or circulatory insufficiency have less effective compensatory mechanisms and that in such cases the individual setting of the ventilator is more important.

6. Application III: A Study on Intermittent Mandatory Ventilation (IMV)

Intermittent mandatory ventilation (IMV) is a ventilatory mode with a complex pattern of breaths. An IMV-cycle begins with a mandatory breath (MB) imposed on the patient by the ventilator, which is followed by a sequence of spontaneous breaths (SB). Since pleural pressure during inspiration rises in the MB and falls in the SB, we assumed that the gas exchange in mandatory breaths is different from that in spontaneous breaths. It is difficult, however, to investigate this hypothesis quantitatively by comparing the mean values of a period of IMV with the mean values of a period of CPPV and of CPAP. Mandatory and spontaneous breaths have to be compared separately. Consequently, we had to use methods suitable for a breath-by-breath analysis. In other words, this study demonstrates special advantages of the measuring system introduced in the present volume. We are able to show that each single breath is investigated separately and the results of the breaths of a certain type – i. e., spontaneous and mandatory breaths – can subsequently be grouped. Moreover, the comparison of IMV with CPAP provides the opportunity to demonstrate the measuring system as it is applied for spontaneous breathing.

IMV was reported to have clinical advantages over other forms of ventilation in that it reduced the need for sedatives, narcotics, and muscle relaxants (Downs 1973, Cullen 1975). While results are available of extensive studies concerning the effects on hemodynamics (reviewed in Weisman 1983), very little quantitative data on gas exchange have been published. Some authors have measured physiological dead space (Downs 1976, Douglas 1980, Hastings 1980, Prakash 1982), respiratory rate, CO_2 elimination, functional shunt fraction, cardiac output by thermodilution (Prakash 1982), and functional residual capacity (Downs 1976). In other studies, mean values obtained over a period of IMV are compared with mean values measured over a period of other modes of ventilation. To our knowledge, there are only two papers that report single breath data on mandatory breath (MB) and spontaneous breaths (SB) (Shutack 1982, Weiss 1983). This lack of data, along with the fact that many IMV systems are not adequately used (Gibney 1982, Kahn 1983, Downs 1980) appears to have caused some controversy with regard to the clinical application of IMV (Weisman 1983, Luce 1981).

The purpose of this study was to investigate patients using quantitative methods appropriate for a breath-by-breath analysis. The comparison of spontaneous breaths and mechanical (mandatory) breaths should provide a better understanding of the IMV mode. It is hypothesized that the quantitative measures support our good clinical experience with IMV. The study has now been published in a slightly abridged form (Wolff 1986). We are grateful to CHEST for allowing us to use the illustrations here.

6.1. Patients and Methods

6.1.1. Patients

We investigated 8 patients following open heart surgery (Appendix 9.3.1) during the weaning period. Immediately after the operation the patients received volume-controlled ventilation (IPPV), and sedation (diazepam), analgesia (morphine), and muscular paralysis (pancuronium) were continued. As soon as intravascular volume, cardiac rhythm (atrial pacing), core temperature, and peripheral circulation were adequate, PEEP was applied (CPPV) and kept constant throughout the subsequent investigations. Inspired oxygen concentration was constant at 40 vol%. After the first set of measurements was performed (CPPV: 6 hours, SD ± 5.7 after the end of the operation), the weaning procedure was introduced by discontinuing muscular paralysis and by reducing the administration of analgesics and sedatives. When the patients started to become active, the ventilator was switched into the IMV mode with identical PEEP. The frequency of the mechanical breaths was slowly reduced and the patients started to breathe spontaneously. Mandatory breaths during IMV were not triggered by the patient. As soon as the breathing pattern was regular and in phase with the ventilator, and the frequency of the mandatory breath was reduced to below 5/min, a second set of measurements was taken (IMV: 10.6 hours [SD ± 5.3] after the end of the operation). As soon as the patients were able to breathe on their own, the ventilator was switched into CPAP mode. When the breathing pattern remained constant, CPAP was investigated (CPAP: 12.5 hours [SD ± 4.9] after the end of the operation). One patient (JCO) was not investigated in CPAP. All of the patients were extubated within 20 hours after the operation and the postoperative course was normal. The protocol of this study was approved by the ethical committee of our hospital.

6.1.2. Measurements

All of the measurements were taken in steady state conditions. Steady state was assumed when the end-tidal CO_2-concentration remained constant for more than 20 minutes. Each set of measurements included a continuous recording of flow, pressure and gas composition over a period of 3 minutes during each mode of ventilation. At the same time, an arterial blood sample was drawn from a systemic artery over a period of at least 20 seconds and was analyzed immediately. Mixed venous blood was sampled simultaneously from the pulmonary artery and its oxygen saturation was determined at bedside. A N_2 washout with argon was then carried out.

During CPPV, IMV, and CPAP, the patients were connected to a modified Monaghan M-250 ventilator. This system is equipped with a highly compliant pressurized inspiratory reservoir. CPAP was maintained by an overflow system (without demand valve), which prevented airway pressure from dropping during spontaneous breathing. Our ventilator enabled us to perform N_2 washouts without producing a change in the ventilatory pattern during CPPV, IMV, and CPAP (Nyffeler 1984).

6.1.3. Presentation of Results

We present the results of single breath analysis as mean values. For CPPV and CPAP, a consecutive series of 15 to 20 breaths was analyzed and the mean values of the breath-by-breath indices are reported. For IMV a consecutive series of 12 to 20 IMV cycles was analyzed. Mean values were determined separately for mandatory breaths (superscript MB). No difference was made between spontaneous breaths within an IMV-cycle, but one mean value was calculated (superscript SB). The number of mandatory breaths and the number of spontaneous breaths was counted and related to one minute, which yielded RR^{SB} and RR^{MB}. The quotient RR^{SB}/RR^{MB} gives the number of spontaneous breaths within an IMV-cycle. For V_T, V_{dS}, alv.eff.-CO_2, and $F_{CO_2\text{-et}}$, we calculated the difference between CPPV and mandatory breaths in IMV, as well as the difference between CPAP and the spontaneous breaths in IMV for each patient (paired differences). All of the spontaneous breaths with tidal volumes of less than 200 ml, of which there were 42, were not included in the breath-by-breath analysis. Statistical analysis was carried out with the Wilcoxon signed rank test for paired samples ($p < 0.05$).

6.2. Results

A summary of the results in CPPV, IMV, and CPAP is presented in the appendix 9, tables 3.2 to 3.4. Single breath indices during IMV (with MB's and SB's separated) are given in Appendix 9, table 3.5.

During IMV, there was a small and non-uniform difference between inspiratory and expiratory volume in the mandatory breaths ($V_I^{MB}-V_E^{MB}$). The series dead space was much smaller in the spontaneous breaths than in the mandatory breaths. The small difference in alveolar ventilation was not uniform and not statistically significant. The most interesting observation was the increase in alveolar efficiency for CO_2 elimination during the spontaneous breaths with a mean increase from 81.2% to 87.4%. One patient (UGE) had a surprisingly large series dead space V_{dS}, which was compensated for by an increased minute ventilation; the same patient showed the largest FRC and the highest shunt fraction.

The breath-by-breath indices were stable in each patient. The mean SEM are given in Table 3.6 of appendix 9. for CPPV, IMV, and CPAP. The largest scatter was observed in the SB's during IMV.

Mean values describing CPPV and CPAP compared with the IMV period are given in Fig. 6.1. In some patients, the FRC was decreased in IMV as compared with CPPV. The difference, however, did not reach the level of significance, whereas the slight reduction of FRC (mean difference = 210 ml) from IMV to CPAP was statistically significant. There were large and significant changes in the CO_2 and mean airway pressure. V'_{CO_2}, MV, and P_{aO_2} changed very little and non-uniformly from CPPV to IMV and from IMV to CPAP.

Some breath-by-breath indices of CPPV, IMV, and CPAP are given in Fig. 6.2. Tidal volumes of the MB's in IMV were similar to those in CPPV and tidal volumes of the SB's in IMV resembled those in CPAP. $F_{CO_2\text{-et}}$ and P_{aCO_2}, on the other hand, were significantly increased in IMV and CPAP as compared with CPPV.

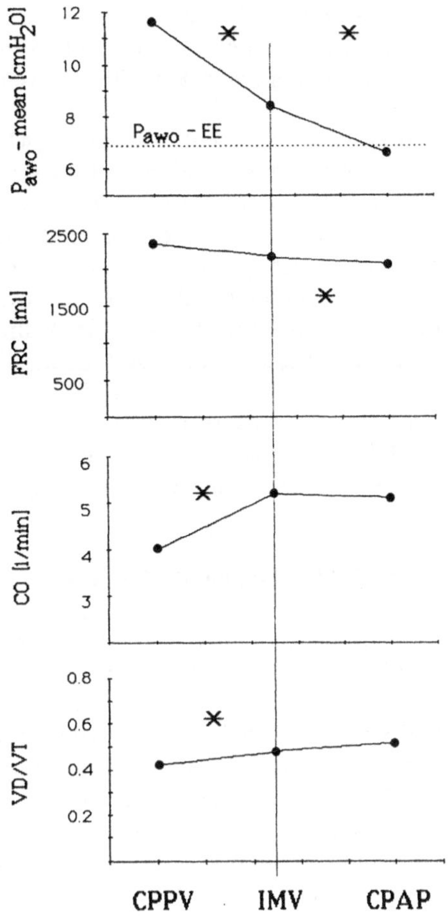

Fig. 6.1. Net effect of gas exchange during periods of CPPV, IMV, and CPAP. Mean airway pressure (P_{awo}), accessible pulmonary gas volume to nitrogen washout (FRC), cardiac output (CO), and V_D/V_T are plotted as mean values. An asterisk indicates statistically significant changes with respect to IMV

From: Wolff G, Brunner JX, Grädel E (1986) Gas Exchange during Mechanical Ventilation and Spontaneous Breathing. Intermittent Mandatory Ventilation after Open Heart Surgery. Chest 89:11–17

6.3. Discussion

Since pleural pressure during inspiration rises in the MB and falls in the SB, we assumed that the gas exchange in mandatory breaths is different from that in spontaneous breaths. It is difficult, however, to investigate this hypothesis quantitatively by comparing the mean values of a period of IMV with the mean values of a period of CPPV and CPAP, respectively. Mandatory and spontaneous breaths have to be compared separately. Consequently, we had to use methods suitable for a breath-by-breath analysis. The disadvantage of these methods is the poor resolution with respect to ventilation-perfusion (V'_A/Q') ratios, which means that we are unable to distinguish between alveolar dead space ventilation and ventilation of areas with poor but finite perfusion, with the exception of the series dead space ventilation. Consequently, we are unable to distinguish between a broad unimodal and a multimodal distribution of ventilation. The same applies to low V'_A/Q' ratios with respect to Q'_S/Q'_T. There are methods which provide better resolution (Wagner 1974) in terms of V'_A/Q' ratios under the assumption of a steady state. These methods, however, are limited with respect to time-resolution. They are neither suited for studying breath-by-breath differences nor for assessing series dead space ventilation.

IMV was investigated an average of 4 hours later than CPPV and therefore the level of anesthesia was reduced when IMV was investigated. In contrast, only an average of 2 hours had passed between the investigations of IMV and CPAP and the level of anesthesia may be considered comparable. Mean body temperature was 37 ± 1.7°C; it increased slightly from CPPV to IMV (mean temp. = 1.3°C) and remained constant thereafter. Therefore, differences in physiological measurements between CPPV and IMV, and perhaps also between IMV and CPAP, are not necessarily due to the mode of ventilation alone. Such a systematic difference in sedation and temperature may be reflected in the low CO during CPPV as compared with IMV and CPAP.

Mean airway pressure did not fall below PEEP in IMV and in CPAP. Obviously, our IMV system adequately maintained the end-expiratory pressure without loading the patient's breathing system with high inspiratory resistance.

6.3.1. CPPV Compared with IMV

No uniform change in CO_2 production was found when switching from CPPV to IMV (nor from IMV to CPAP). In another study (Prakash 1985) on patients similar to ours, but in which an IMV system with demand values was used, CO_2 production during CPPV was found to be similar to that in our study but increased by 40% during IMV. In that study, the ratio of CO_2 elimination and O_2 uptake (RQ) was found to be 1.03 in CPPV and to decrease to 0.68–0.77 during IMV, a point which the authors did not mention in the discussion of their results. In our study, RQ was about 0.95 during CPPV and did not decrease during IMV or CPAP. We are reluctant to attribute these discrepancies to differences among the patients. We rather believe that they result from technical difficulties caused by the sensors being placed at various sites, i.e., not all of them being located directly at the airway opening. We consider the measurement of gas flow and gas concentrations directly at the airway opening mandatory for obtaining accurate results during IMV.

With the exception of one patient (ZGR), mean airway pressure was reduced in IMV. This result is plausible, since the thorax in IMV is no longer passively inflated. Gas flow into the lungs (at least during the SB's) is caused by a lowering of the alveolar pressure, which thereby reduces the mean pressure at the airway opening.

In IMV we measured a slight increase in V_D/V_T (Fig. 6.1), a result which is contradicted by some authors (Hastings 1980, Prakash 1982). Physiological dead space ventilation (V_D/V_T) is composed of series dead space ventilation and alveolar dead space ventilation (measured as the inverse of alv.eff.-CO_2). Alveolar dead space ventilation as well as V_{ds} are decreased during IMV (Fig. 6.2). The small spontaneous volumes during IMV, however, show a higher V_{ds} to V_T ratio. The series dead space ventilation (in terms of ml/min) is thus increased, despite decreased V_{ds} (in terms of ml/breaths). Our conclusion is that the slightly increased V_D/V_T observed in our patients during IMV is caused by an increased series dead space ventilation per minute and not by an increased alveolar dead space ventilation. This finding is

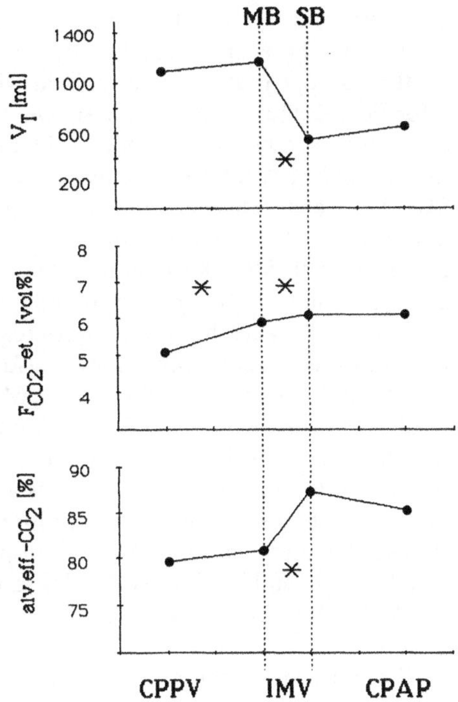

Fig. 6.2. Breath-by-breath indices of gas exchange during CPPV, IMV, and CPAP. Mean values of tidal volume (V_T), end-tidal $CO_{2\text{-et}}$) and efficiency of CO_2-elimination (alv.eff.-CO_2) are plotted in CPPV, the mandatory breaths in IMV (MB), the spontaneous breaths in IMV (SB) and CPAP. An asterisk indicates statistically significant differences

From: Wolff G, Brunner JX, Grädel E (1986) Gas Exchange during Mechanical Ventilation and Spontaneous Breathing. Intermittent Mandatory Ventilation after Open Heart Surgery. Chest 89:11–17

dependent on alterations in V_{ds}, alv.eff.-CO_2, and the ratio RR^{SB}/RR^{MB}. Consequently, V_D/V_T depends on both the patient and the setting of the ventilator.

6.3.2. IMV^{MB} Compared with IMV^{SB}

IMV is composed of mandatory breaths and spontaneous breaths (Fig. 6.2, Appendix 9, Table 3.5). They differ with regard to two major features: the tidal volumes of MB's are larger than those of the SB's and pleural pressure rises during mandatory inspiration while it falls during spontaneous inspiration.

The inspired and expired volume of the MB's did not differ. The same was true for the SB's. Consequently, end-expiratory lung volume or functional residual capacity was constant throughout the IMV-cycle.

Series dead space volume (V_{ds}) is much larger in MB than in SB (mean = 59 ml). V_{ds} develops during inspiration and three mechanisms could contribute to this alteration. It has been known for some time that V_{ds} is increased if end-inspiratory lung volume ($V_{L\text{-}EI}$) is increased (Martin 1979). The tidal volumes of the spontaneous breaths were only 46% (mean) as compared with the tidal volumes of the mandatory breaths, which results in a larger end-inspiratory lung volume in the MB. Therefore, one would expect V_{ds} to be larger in the MB. In the study mentioned (Martin 1979), however, V_{ds} $V_{L\text{-}EI}$ is 25 ml/l. In our study, the (mean) difference in end-inspiratory lung volume after the SB's in comparison with after the MB's is 620 ml, which explains a difference in V_{ds} of only 15 ml, i.e., our results cannot be fully explained by this mechanism. Two further mechanisms have to be discussed, both of which are based on the concept that V_{ds} is contained within "anatomical" and "functional walls" (Horsfield 1968). It has been shown that during positive pressure ventilation, convective airways are distended (Hedenstierna 1975a). We can therefore assume that during IMV, convective airways are distended more in the MB's and less in the SB's. Since convective airways are the "anatomical walls" of V_{ds} we conclude that the distensibility of the airway is the second mechanism of the increase in V_{ds} in MB as compared with SB. The stationary interfaces (SI) between fresh gas and "alveolar gas" (Cumming 1971) are considered "functional walls" of V_{ds} (Wolff 1984). The position of SI may be moved upwards (towards the mouth) by decreasing the inspiratory convective flow (Paiva 1972) as well as by an end-inspiratory pause (Martin 1979). Inspiratory flow during MB is constant (rectangular) up to the end of inspiration, the transition of inspiration to expiration being very fast (0.1 sec). On the other hand, in the SB's, inspiratory flow comes down slowly to zero. Since the position of the SI's depends on the inspiratory flow pattern, particularly at the end of inspiration, we conclude that the different breathing patterns cause the difference in V_{ds}. All three mechanisms (end-inspiratory lung volume, end-inspiratory airway pressure, and inspiratory flow pattern) reduce V_{ds} in the SB's as compared with the MB's. We are not able, however, to assess the magnitude of the individual effects.

The end-tidal concentration of CO_2 ($F_{CO_2\text{-et}}$) increased slightly in the SB's (mean = 0.2%) and the efficiency for the alveolar CO_2 elimination (alv.eff.-CO_2) was far greater than that in the MB. To explain these observations we should discuss three possible mechanisms:

1. alveolar ventilation is decreased in the MB;
2. dead space ventilation is increased in the MB;
3. pulmonary perfusion is decreased in the MB.

The change in alveolar ventilation is very small and non-uniform (Table 3.5 in appendix 9) and therefore most probably is not the explanation of our results. It has been shown that in supine, spontaneously breathing patients, inspired gas is preferentially distributed to dependent lung regions (Froese 1974). In contrast, during CPPV, ventilation is directed to the non-dependent lung regions (Rehder 1975). Although these effects are mixed during IMV, they may lead to large differences in ventilation and perfusion between spontaneous breaths and mandatory breaths. We measured an increased alv.eff.-CO_2 in the spontaneous breaths as compared with the mandatory breaths in each patient. This finding suggests that ventilation in high V'_A/Q' regions in the MB is further increased and that the results just mentioned (Froese 1974, Rehder 1975) may be important, even within an IMV cycle. It is not clear what the diaphragm contributes to this difference. Since the diaphragm is usually active in the SB's, it is highly unlikely that no diaphragmatic muscular action takes place in the MB.

The displacement of blood from the thorax during positive pressure breathing has been recognized for many years (Fenn 1947). Recently, it has been demonstrated that during positive pressure ventilation, pulmonary capillary perfusion is reduced at the end of inspiration (Pinsky 1984). Thus, V'_A/Q' may vary, even within a respiratory cycle. This tidal (= temporal) change in the V'_A/Q' was considered to contribute to the increased V_D/V_T during mechanical ventilation (Cooper 1961). Our results support the hypothesis that such effects also take place within the MB during IMV.

Our results suggest that during the spontaneous breaths either capillary perfusion in non-dependent regions is increased or ventilation is redistributed towards regions of lower V'_A/Q' ratios, or both.

6.3.3. IMV Compared with CPAP (Figs. 6.1 and 6.2)

CO_2 production as well as cardiac output and $a\bar{v}D_{O_2}$ remained constant, although in CPAP the patients had to breathe totally on their own. Therefore, the mandatory breath during IMV does not necessarily reduce the amount of a patient's work due to breathing, provided an adequate IMV system is used. FRC was decreased in CPAP. There was no significant change in arterial oxygenation and Q'_S/Q'_T did not change. Therefore, in situations where a reduced FRC could cause clinical complications, IMV might be an advantage over CPAP.

Nota Bene:

- The breathing pattern in IMV can be broken down into cycles. Each IMV cycle begins with a mandatory breath (MB), which is followed by several spontaneous breaths (SB's).
- IMV can be investigated by methods with a time resolution that allows breath-by-breath measurements.
- During IMV, alv.eff.-CO_2 is always much greater in SB's than in the MB. This observation supports the hypothesis that the match between perfusion and ventilation during SB's is better than during MB.
- There is some controversy in the literature over the effect of IMV on V_D/V_T. However, V_D/V_T is composed of series dead space ventilation and alveolar dead space ventilation and changes of V_{ds} and alv.eff.-CO_2 may depend on different mechanisms. We conclude that V_D/V_T is not a useful index and should not be used to describe IMV.
- O_2-uptake and CO_2 elimination was not reduced when IMV was changed to CPAP. We conclude that even with good equipment IMV (as compared with CPAP) does not necessarily reduce the work of breathing.
- We have had excellent clinical experience with the IMV system used in this study. We have the impression that IMV makes it possible to start spontaneous breathing at a time the patient is still unable to maintain adequate blood gases on his own.
- Our study demonstrates that in this situation mean airway pressure can be reduced with a constant FRC, with constant oxygenation, and with less dead space ventilation.

Appendix 1

On the Morphology of the Lungs

1. Upper Airways

The airways (nose-trachea-lungs) and the digestive tract (mouth-esophagus-stomach) cross paths in the throat (larynx). By upper airways is meant the space above the larynx. Located here are the nasal cavity and pharynx. In addition, the oral cavity is connected (end-to-side) to the upper airways, but gas flow through the mouth is usually prevented by the soft palate. Many persons breath through the mouth when large tidal volumes with high rates are necessary (or if the nose is closed by swollen mucous layers). In such a situation, flow through the nasal cavity is usually blocked by the soft palate, which is pressed against the hard palate. The same applies during a pulmonary function investigation when a person is breathing via an oral mouthpiece. Producing a model of this area has proved to be a difficult venture (Proctor 1964). Recent investigations using ultrasonic measurements give a more detailed description (Fredberg 1980). Even under normal physiological conditions, slight deformations can occur as a result of movements of the tongue and lower jaw, rapid gas flow, and articulation.

In an adult person (without endotracheal intubation) the space between the lips and the vocal cords (glottis) may be modelled as a tube of 17 cm in length with a volume of 50 ml.

2. Lower Airways

The lower airways are located between the larynx (proximally) and the terminal bronchioles (distally) with a diameter of approx. 0.7 mm (Horsfield 1968). The larynx is the voice organ; the vocal cords are a prerequisite for the coughing mechanism and enables the airways to close when one swallows. The trachea consists of concentric horseshoe-shaped bridges of cartilage which are connected by concentric strips of muscle fiber and connective tissue. The inner surface of the lower airways is lined by mucous membrane ciliated epithelium.

The space between the vocal cords and the first branching of the trachea ("bifurcation" or "carina", which is the origin of the right and left main bronchus) may be described as a tube having a length of 12 cm and a volume of 30 ml.

The bronchial tree with its numerous branches is an extremely complex structure. Quantitative assessment of such a structure is often performed using resin casts. However, this technique requires an enormous amount of time and the number of lungs that can be investigated is therefore limited. Although the lungs are built according to structural principles, each lung in detail has its own individual structure (as, for instance, an elm tree is distinct from other trees). Consequently, the results of such measurements can only be described statistically (number and angle of the branches, length and diameter, etc.).

3. The Intubated Patient

In the ventilated patient the upper airways are bypassed by the endotracheal tube. Usually the endotracheal tube is inserted via the mouth, rarely via the nose. The commercially available tube is usually shortened to about 27 cm. The diameter of the tube depends on the size of the patient's glottis and trachea. In the adult patient an inner diameter of 8 mm (7.5–8.5) is mostly used. The tracheal tube ends some centimeters above the carina, i. e., approximately after bypassing half the length of the trachea. In the intubated patient, the upper airways and the lower airways are therefore not separated.

The endotracheal tube and the free part of the trachea may be compared with a tube having a length of about 33 cm and a volume of about 70 ml.

4. Conductive Airways

The total tubing system between the airway opening (lips, nostrils, or opening of endotracheal tube) and the terminal bronchioles is called the conductive airways or convective airways.

The lower airways can be modelled because they are not as susceptible to changes in shape (the shape does in fact change, but the structure remains the same) as the upper airways. Weibel (1963) describes the bronchial system as a tree in which every branch further divides into two "daughter" branches (dichotomy). This dichotomy, however, is irregular and the two daughter branches can differ in length and diameter. Weibel presented two classification models: Weibel model A with regular dichotomy, and Weibel model B with irregular dichotomy. The latter is much more realistic. Fig. A 1.1 shows the principle set up of Weibel model A. The branches are numbered starting with the trachea as 0 and increasing by 1 after each bifurcation, with the two daughter branches both given the same number. These numbers are called "generation numbers".

A total of 16 different generations of conducting airways result, ranging from diameters of 18 mm in the trachea to 0.6 mm in the terminal bronchioles (generation 16). Respiratory bronchioles carry spherical protrusions (alveoli) in the walls of which (alveolar membrane) a dense network of blood capillaries runs. The exchange of O_2 (from the alveolar gas into the blood) and CO_2 (from the blood into the alveolar gas) takes place through the alveolar membranes.

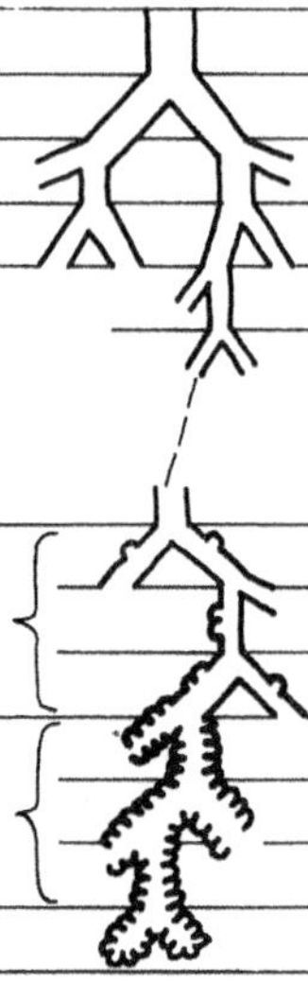

	Generation		Diameter [cm]	Length [cm]	Number	Total cross-section [cm²]
		0	1.80	12.0	1	2.54
		1	1.22	4.8	2	2.33
		2	0.83	1.9	4	2.13
		3	0.56	0.8	8	2.00
		4	0.45	1.3	16	2.48
		5	0.35	1.07	32	3.11
		↓	↓	↓	↓	↓
		16	0.06	0.17	6×10^4	180.0
respiratory zone		17	↓	↓	↓	↓
		18				
		19	0.05	0.10	5×10^5	10^3
	T_3	20	↓	↓	↓	↓
	T_2	21				
	T_1	22				
	T	23	0.04	0.05	8×10^6	10^4

Fig. A 1.1. Weibel's model of the lungs (from Weibel 1963, slightly modified). The trachea has the number 0, the first respiratory bronchiole 17, and the alveolar ducts 20 to 23

Horsfield and his assistants in a tedious undertaking demanding enormous patience numbered resin casts of human lungs. They selected the bronchioles of 0.7 mm in diameter as the smallest unit and counted 17 generations up to the trachea. Fig. A 1.3 gives a summary of some of the bronchial tree measurements along with analog measurements made on the pulmonary-arterial system.

Strahler's system (Woldenberg 1970) Fig A 1.2 is better than Weibel's for grouping together small structures of the same morphology and function. Unfortunately, however, it cannot be applied in everyday clinical practice. For bronchoscopy or surgery, for example, Weibel's numbering system is still the most useful.

5. Lung and Respiratory Zone

The lungs consist of the right and the left lung. The right lung can be further divided into 3 lobes with 10 segments. The left lung has 2 lobes and 8 segments. The entire structure is invested by the visceral pleura, which is separated from the parietal pleura (covering of chest wall and diaphragm) by a lubricating secretion. The "non-existing", i. e., virtual space between the visceral pleura and parietal pleura is called the pleural space.

The respiratory zone begins where the first alveoli appear – with the respiratory bronchioles. The respiratory bronchioles are the origin of several generations of alveolar ducts (AD), whose walls consist completely of alveoli. These ducts lead to the alveolar sacs (AS). The alveoli themselves are described as a cluster of round structures. Many observations demonstrate that gas can be exchanged between

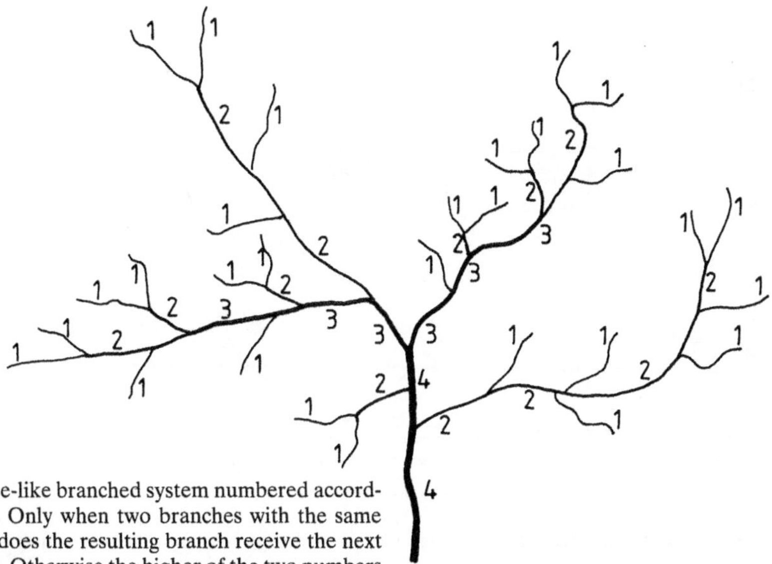

Fig. A 1.2. Tree-like branched system numbered according to Strahler. Only when two branches with the same number merge does the resulting branch receive the next highest number. Otherwise the higher of the two numbers is retained

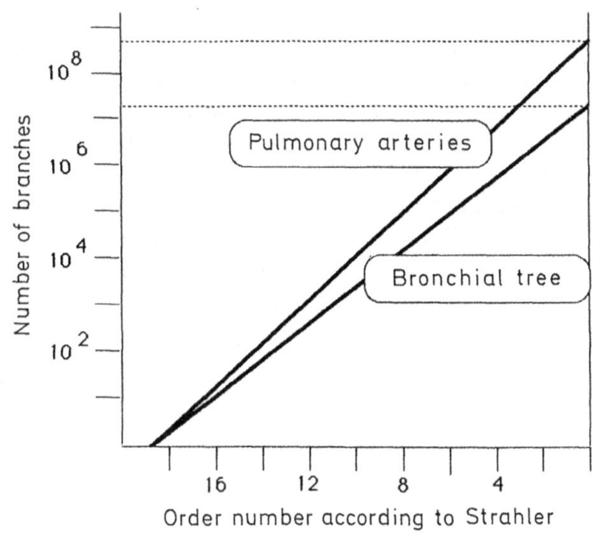

Fig. A 1.3. Characteristics of the branching of the pulmonary arterial capillary network and the bronchial system. In the bronchial system each branch is the origin of an average of 2.74 daughter branches. The pulmonary vascular system in contrast is increased 3.4 times with every branching (Cumming 1977)

neighboring alveoli, even when the supplying bronchiole has been blocked. This phenomenon is called collateral ventilation. Between the adjacent alveoli are direct communications known as pores of Kohn. The function of these pores is not exactly known. They are thought to serve as a means for collateral ventilation.

The respiratory unit, also called the acinus, is defined differently by various authors. Some define the origin of the acinus as the respiratory bronchioles, others claim it to be the terminal bronchioles. We use the second definition. Weibel as well as

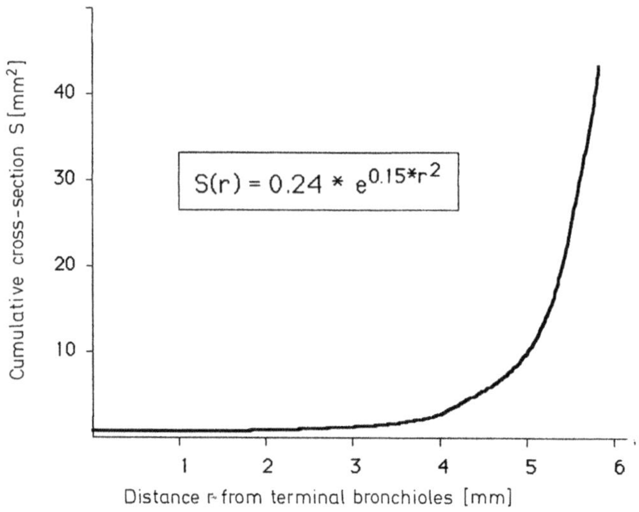

Fig. A 1.4. Cumulative cross-section in acinus as function of the distance from the entrance

Horsfield and Cumming found an enormous increase in the cumulative cross-sectional area in the acinus (from 300 cm^2 in generation 17 to 11800 cm^2 in generation 23 according to Weibel). Hansen and Ampaya (1975) described for an average acinus a correlation between the cross-sectional area (S) and the distance (r) from the terminal bronchiole (eq. in Fig. A 1.4; unit in mm). Fig. A 1.4 illustrates the increase in the cross-sectional area.

Rotating the graph in Fig. A 1.4 around the abscissa produces a spherical body that closely resembles a thumbtack. LaForce and Lewis (1970) therefore gave it the name "thumbtack model". This structure has major consequences for gas transport and gas mixing in the lungs (see Appendix 4).

Appendix 2

Technical Principles of Mechanical Ventilation

Following the poliomyelitis epidemic in the fifties, intermittent positive pressure ventilation rapidly became adopted to substitute for the paralyzed muscles. It also became employed more and more to treat pulmonary respiratory insufficiency (Lassen H.C.A. 1953). This resulted in numerous ventilators being built. The principle tasks of a ventilator are:

1. to supply the patient with enough fresh gas;
2. to produce conditions under which a pulmonary disease can heal (ventilatory variables);
3. to prevent conditions which would bring on complications (e.g. barotrauma).

To fulfil these tasks the ventilator must provide a source of gas, on the one hand, and on the other, *control* the flow of gas to ensure that the patient receives the desired minute volume. To do this, inspiration (gas source active) and expiration (expiration valve opened totally or partially) take turns in cycles, which are controlled by the ventilator (see Fig. 1.1). The control of inspiration depends on the mode of breathing and, since older type ventilators only have a few modes of operations, is sometimes used as a means to classify ventilators.

1. Controlled Ventilation

Controlled ventilation means that the beginning of inspiration and expiration as well as their duration are automatically determined by the ventilator. The ventilator carries out complete control of the breathing.

1.1. Control of the Gas Flow

There are two different kinds of gas sources: a flow source and a pressure source. The atmosphere, for instance, provides healthy people with an ideal source of totally adequate pressure. The wall outlets in hospitals, however, offer only highly pressurized air. The primary task of a ventilator therefore is to reduce that pressure to a physiologically acceptable level. This is often accomplished by a variable resistor, which basically turns the high pressure source into a variable flow source (e.g. Bird ventilator). Or one can separate the patient from the high pressure source by means of valves and bellows. This works as follows: during expiration, the bellows are connected to the high pressure gas source and are filled to an adjustable level. During the subsequent inspiration, the bellows are disconnected from the gas source and connected to the patient, where they then empty into the patient. The driving forces are the recoil forces of the bellows or the static energy of a weight lifted during expiration (e.g. Bromptom Main Ventilator) or externally applied pressure (e.g. Monaghan M-250). The third technical solution uses closed loop control. The physician sets the desired values for either flow or pressure source. The ventilator measures its output with a sensor, compares it with its current settings, and adjusts itself to achieve the output set by the physician. If the sensor measures flow, the ventilator becomes a source of flow; if it measures pressure it becomes a source of pressure (e.g. Siemens 900 C).

1.2. Control of Valves

Pressure-limited inspiration
When the ventilator delivers a breath to the patient, pressure builds up at the mouth. The magnitude of the pressure rise is dependent on gas flow, volume, resistance, total compliance, and the patient's breathing movements (see Fig. 1.2). This pressure is measured in the ventilator and the expiration valve is opened as soon as a certain value has been exceeded. Since by his breathing movements the patient actively takes part in determining the pressure, he has a certain amount of control over the ventilator, though only in a negative sense. The patient is able to reduce the single breath volume, the minute volume decreases and the patient can be hypoventilated (example: Bird ventilator).

Volume-controlled inspiration
This mode of breathing is usually used together with a constant flow gas source. Less frequently used flow patterns are sinusoidal, ascending, and descending inspiratory flow patterns. A machine that performs volume-controlled ventilation acts like an air syringe. The breath volume set is pressed into the lungs without any consideration of the pressures that form. This type of ventilation ensures delivery of the set MV (Example: Monaghan M-250).

Time-limited inspiration
The duration of inspiration is determined by a timer circuit. When, for instance, there is constant inspiratory gas flow to the patient, the breath volume is limited by the inspiration time set (example: Megamed Assistomat).

2. Assisted Ventilation

This mode of breathing assumes that the patient is able to breathe spontaneously. It does not mean that the patient has to provide the entire minute volume himself.

However, a certain if minimal amount of activity on the part of the patient is required when this mode of breathing is used.

Modern ventilators offer a number of new and more sophisticated modes of ventilation or a combination of spontaneous breathing with some sort of support and mechanical ventilation. These modes include synchronized intermittent mandatory ventilation SIMV, minimal minute ventilation MMV, pressure supported ventilation, and many others. The performance of these modes and the benefit for the patient depend very much on the technical solution i.e., on the particular ventilator used (see chapter on IMV). The large variety precludes our describing them all. We do, however, encourage the reader to study these modes of ventilation thoroughly before judging the ventilator. A help in this respect is the discussion of the following question:

Is a particular mode available because of its potential benefit for the patient or is it included because the ventilator performs well in this mode?

2.1. Control of Gas Flow

Assisted ventilation is usually used together with an elevated airway pressure, also during expiration. The result is that, ideally, the pressure at the mouth never decreases to atmospheric pressure P_B (continuous positive airway pressure CPAP), but is kept above it. The pressure difference is called the end-expiratory pressure or PEEP.

In theory, the ventilator provides the patient with an ideal pressure source to maintain the set PEEP. In practice, however, the PEEP level is reached with three different methods:

1. Expiration is stopped as soon as the airway pressure falls to below a set level (PEEP) (e.g. Monaghan).
2. A constant gas flow through a T-piece at the mouth causes a pressure drop over the expiratory limb (PEEP), which is proportional to the flow and the resistance of the expiratory limb.
3. Pressure at the mouth is maintained by closed loop control using a pressure sensor at the mouth. Unfortunately, most ventilators have their pressure sensor in the machine and not at the mouth (e.g. Siemens).

When properly applied, solution # 2 is still the best technique (see IMV study).

2.2. Control of valves

The patient himself starts the beginning of inspiration by inspiratory movements, i.e., by reducing the intrathoracic pressure and consequently, the airway pressure. In other words, the patient triggers off the machine. The patient influences how long expiration lasts and how great the inspired volume is. However, when the maximum permissable expiration time has been exceeded, the next inspiration is automatically started by the ventilator thereby maintaining a minimal ventilation (safety feature of most ventilators).

Appendix 3

Measuring the Pleural Pressure with an Esophageal Balloon

The use of an esophageal balloon to measure the pleural pressure in spontaneously breathing patients has been adequately studied, widely adopted, and standardized. There are no data, however, on its use for mechanically ventilated patients. This is mainly because the pleural pressure (P_{pl}) and esophageal pressure (P_{es}) can only be compared during occlusion tests (Milic-Emili 1964). And occlusion tests require intact spontaneous breathing.

Occlusion Test

The occlusion test requires that the volunteer (or the patient) produces spontaneous inspiratory and expiratory breathing movements while keeping the glottis open but the mouth closed (occlusion). This causes the pleural pressure to drop and rise and the alveolar pressure (P_A) along with it. The lung volume remains constant as a result of the occlusion (compression is not taken into account) and thus $P_{pl} = P_A$ applies. Since there is no gas flowing, no pressure gradient can form along the airways and the pressure measured directly at the mouth (P_{awo}) can be assumed to be identical with the (non-measureable) alveolar pressure. Provided the lungs expand and compress uniformly and, above all, that the breathing pattern of the maneuver is the same as the breathing pattern before and after the occlusion test, the pleural pressure measurement in the esophagus can be checked for validity by comparing the P_{awo} and the P_{es}. In other words, if the pleural pressure measurement is valid, the signals of P_{es} and P_{awo} are identical during the occlusion test.

Higgs (1983) and Drummond (1983) carried out occlusion tests (OT) in spontaneously breathing patients in a supine position under anesthesia. However, to perform a similar test in paralyzed and therefore mechanically ventilated patients, we had to use a special maneuver. This was called the abdominal pressure maneuver (APM). It involves closing the endotracheal tube at the end of an expiration and within the next 3–5 seconds slowly pressing the abdomen with the palms of the hands until the diaphragm is pushed upward and the intrathoracic pressure increases. Then the hands are slowly removed from the abdomen and the pressure drops again within the next few seconds. Finally, the endotracheal tube is reopened. The maneuver is repeated several times so that the total apnea lasts a maximum of 20 seconds.

Investigations

Five patients were investigated following heart surgery (age: 57.8 ± 5.3 years; weight: 76.2 ± 3.6 kg; height: 173 ± 6 cm). The probe used was a stomach probe with an integrated esophageal balloon (Malinckrodt, 10 ml balloon having a length of 10 cm and 4 openings at distances of 2.5 cm in the pressure measurement channel). It was refrigerated prior to the induction of anesthesia for it to stiffen and was advanced via the nose far enough to ensure that it had safely reached the stomach (70 cm from the nostrils). Before we began our postoperative investigations, the probe was pulled back until the balloon was about 45 cm from the nostrils. Any residual gas was withdrawn with a syringe and the balloon was then filled with approximately 0.8 ml air.

In four patients the abdomen pressure maneuver (APM) was carried out during IPPV immediately after surgery. In three patients the occlusion test (OT) could be performed during CPAP in the later weaning phase. P_{awo} and P_{es} were stored on disk via an A/D converter and analysis was done as described in the following.

Data Analysis

The A/D converter has a resolution of 4095 units (ADU). Of these, approximately half (2000 ADU) were used. The P_{awo} values were divided in steps of 10 ADU and the simultaneously measured values of the P_{es} were averaged separately in each of the groups of 10 steps for each patient and each maneuver. The purpose of establishing the mean value was to eliminate cardiogenic oscillations. Seven curves resulted which were evaluated with respect to linearity (r^2) and proportionality ($k = P_{es}/P_{awo}$) by means of linear regression.

Results

The pressure amplitudes generated by the APM were 13.6 ± 1.4 cmH_2O and those produced by the OT were − 35 ± 1 cmH_2O (inspiration) and 12 ± 8 cmH_2O (expiration). The results of the regression analysis are given in Table A2.

Table A.2 Relationship between P_{awo} and P_{es} during Abdominal Pressure Maneuver (APM) and Occlusion Test (OT); $k = \Delta P_{es}/\Delta P_{awo}$

Patient	k_{APM}	r^2_{APM}	k_{OT}	r^2_{OT}
Ae	–	–	1.01	0.995
Am	0.713	0.968	0.82	0.992
Fu	0.84	0.996	0.61	0.968
Ki	0.727	0.985	–	–
We	0.636	0.990	–	–
mean	0.73		0.81	

Discussion

Occlusion Test

Although the catheter used was slightly different from the classical model (stiff and open at the end), the results of the OT are nevertheless comparable with those obtained by Higgs (1983), who used the classical catheter. The relationship between P_{awo} and P_{es} is linear throughout the entire range (see correlation coefficients in Table A2). This suggests that the relative changes can be accurately measured, while the P_{es} does not reflect the P_{pl} at a 1:1 ratio (mean value 1:0.81). However, prerequisite for this conclusion is that the results be applicable to spontaneous breathing. Drummond (1983) presented a thorough discussion of these problems, which include compression and expansion of the intrapulmonary gas volume, the possible redistribution of the gas volume, phase shifting, particularly in the case of chronic obstructive lung disease, and inhomogenous compression and expansion caused by the action of the various muscle groups. The last effect mentioned is discussed in the following (see Drummond for the others).

The changes in pleural pressure during breathing are not uniform within the pleura. Particularly when only one group of muscles is active, the P_{es} does not reflect a "mean" pleural pressure. Activities of the diaphragm generate greater pressure differences in the diaphragmatic pleura and intercostal activity produces greater pressure changes in the costal part (D'Angelo 1974). We can assume, especially in the patients investigated here after thoracotomy, that the breathing movements during the OT mainly take place via the diaphragm, which causes severe inhomogenities of the pleural pressure.

Both Drummond and Higgs established that the quotients P_{es}/P_{awo} were dependent on the position of the balloon and the local pressure distribution. Higgs suggested changing the balloon position until the ratio P_{es}/P_{awo} in the OT became 1. This suggestion was based on the finding that the linearity between P_{es} and P_{awo} was maintained in every balloon position. Assuming this, a mathematical compensation also suffices, provided the factor is known. The disadvantage of Higgs' suggestion: when a change in the type of breathing affects the pleural pressure distribution, this factor (or the optimal position of the balloon) probably changes as well. Higgs's suggestion therefore does not offer a solution to the practical problems associated with the intensive care patient. And Drummond then concluded that the P_{es} would have only very limited value as a substitute for a mean P_{pl}.

Abdominal pressure maneuver

During IPPV and CPPV the conditions are different. In the patient with muscle paralysis the lungs are distended independently of any muscle groups. During APM, only the diaphragm is moved, which provokes a considerable inhomogeneity. That means that the situation during APM is not the same as that under mechanical ventilation and therefore the maneuver is not a suitable method for finding the ratio P_{es}/P_{awo} during ventilation. And Higgs's suggestion cannot correct this. Since, however, it can be expected that the lungs are distended more uniformly during ventilation than during the APM, the changes in P_{es} should then correspond more to those of P_{pl}. The factors k_{APM} in Table A2 are therefore to be considered minimum values. During ventilation, they are expected to be closer to 1.

Conclusions

- Absolute values of P_{es} may not be interpreted in relation to P_{pl}. If P_{es} is used to calculate the lung distensibility C_L, its absolute value must be interpreted with caution.
- Relative changes in C_L can be determined under the assumption that the pressure distribution in the pleura has not changed.
- P_{es}/P_{awo} measurements during an abdominal pressure maneuver are not applicable for IPPV and CPPV. During mechanical ventilation, however, P_{es} and P_{pl} are more likely to agree than during the abdominal pressure maneuver.

Appendix 4

Transport Equation for Convection and Diffusion

The Stationary Interface

In this section we are dealing with one-dimensional gas flow (in the direction of r) in time invariant geometry. The gas consists of components (n), given by their concentration (C_n). In physiology, the mass of the dry respiratory gas is expressed in volume under standard temperature and pressure and the concentration is given as fractional concentration of dry gas (Piiper 1975). In this chapter however, concentration is meant the mass of a gas per volume.

The following abbreviations are used. For purposes of simplification, the functional dependencies are omitted in the text.

$C_n(r,t)$: concentration of a gas n
$v(r,t)$: convection velocity
$M_n(r,t)$: mass of a gas n
$A(r)$: cross-sectional area
$S(r)$: cross-sectional area in acinus including alveoli
$s(r)$: cross-sectional area in acinus excluding alveoli
$r(t)$: axial distance
D: diffusion constant

The mass transport of a gas by convection (M'_{nK}) can be described by [A4.1]:

$$M'_{nK} = A \cdot v \cdot C_n \qquad [A4.1]$$

Mass transport of gas n by diffusion (M'_{nD}) is based on concentration gradients (eq. A4.2). The total mass transport M'_n is obtained by superimposing convection and diffusion (eq. A4.3).

$$M'_{nD} = -A \cdot D \cdot \delta C_n/\delta r \qquad [A4.2]$$
$$M'_n = M'_{nK} + M'_{nD} \qquad [A4.3]$$

The law of conservation of mass can be formulated in a one-dimensional system by equation A4.4:

$$\delta(A \cdot C_n)/\delta t + \delta M'_n/\delta r = 0 \qquad [A4.4]$$

The cross-sectional area is time invariant ($\delta A/\delta t = 0$) per definition, and the transport equation for convection and diffusion is obtained by the substitution of M'_n in [A4.4] with eq. A4.3 (Paiva 1972):

$$A \cdot \delta C_n/\delta t + \delta(A \cdot v \cdot C_n)\, \delta r - \delta\, [A \cdot D \cdot \delta C_n/\delta r]/\delta r = 0 \qquad [A4.5]$$

Any acinus model (e.g. a cylinder) can now be inserted for A(r). A general model is described in the following.

The first assumption is that there is only a diffusion limitation within the alveolar ducts, i.e., immediate concentration equilibration is assumed within one alveolus (Paiva 1972). The model of the lungs thus breaks up into two parts: the volume in the alveolar ducts and that in the alveoli (Fig. 1 in Paiva 1976). The cumulative cross-section of the alveolar ducts are given by the function s(r), that of the entire acinus (including alveoli) by S(r). Going under this assumption, equation A4.5 can be represented by superposition with the aid of the two partial volumes s(r) (diffusion only) and S(r) (convection only):

$$S \cdot \delta C_n/\delta t + \delta(S \cdot v \cdot C_n)/\delta r - \delta[s \cdot D \cdot \delta C_n/\delta r]/\delta r = 0 \qquad [A4.6]$$

and from this follows

$$\delta C_n/\delta t = D \cdot s/S \cdot [\delta^2 C_n/\delta r^2 + 1/s \cdot ds/dr \cdot \delta C_n/\delta r] - 1/S \cdot \delta(C_n \cdot S \cdot v)/\delta r \qquad [A4.7]$$

Under the assumption that the flow in the system is constant ($S \cdot v = V'$) and that there are steady state conditions ($\delta C_n/\delta t = 0$), the inflection point of the concentration profile ($\delta^2 C_n/\delta r^2 = 0$) results according to eq. A4.8:

$$ds/dr = V'/D \qquad [A4.8]$$

The location of the stationary interface is hence determined by the geometry s(r), the inspiratory gas flow V′, and the diffusion constant D. It is *independent* of the concentration gradients and of the inflowing volume.

Appendix 5

Relationship Between Dead Space and V'_A/Q' Scatter

In this section we intend to show with simple mathematics that V'_A/Q' scatter and dead space ventilation result in the same impairment of gas exchange, i.e., are functionally equivalent.

Fig. 2.18 illustrates an example of an inhomogenous V'_A/Q' distribution. For purposes of simplification a bimodal distribution without scatter is assumed, i.e., two compartments with different V'_A/Q' ratio. A detailed model analysis can be found in West (1969). It is assumed that the model does not possess any series dead space. In the model in Fig. 2.18, the following applies for CO_2:

$$P_{c_1} = P_{A_1} \text{ and } P_{c_2} = P_{A_2} \text{ and } P_v > P_a \text{ and } P_{c_1} \neq P_{c_2} \text{ or}$$
$$F_{c_1} = F_{A_1} \text{ and } F_{c_2} = F_{A_2}$$
(F's represent the fractional concentrations of dry gas)

For further calculation it is assumed that V'_{A_1}/Q'_{c_1} is greater than V'_{A_2}/Q'_{c_2}, but that the two compartments are perfused identically ($Q'_{c_1} = Q'_{c_2} = Q'/2$).

From this it follows that

$$P_{c_1} = P_{A_1} < P_{A_2} = P_{c_2} \quad \text{and} \qquad [A5.1]$$
$$V'_{A_1} > V'_{A_2} \qquad [A5.2]$$

In steady state the amount of CO_2 that is transported from the blood into the lungs is the same as the amount of CO_2 that is eliminated from the lungs, i.e. expired. If the inspiratory CO_2-concentration is zero, the following holds for the CO_2 volume eliminated per minute (V'_{CO_2}):

$$V'_{CO_2} = V'_{A_1} \cdot F_{A_1} + V'_{A_2} \cdot F_{A_2} \qquad [A5.3]$$
$$V'_{CO_2} = Q_{c_1} \cdot (C_v - C_{c_1}) + Q_{c_2} \cdot (C_v - C_{c_2}) \qquad [A5.4]$$

where C is the content of CO_2 in the blood. The dissociation curve for CO_2 can be linearized for purposes of a rough estimation so equation A5.5 applies for the relationship between CO_2 partial pressure P and the CO_2 content:

$$C = \beta \cdot P \qquad [A5.5]$$

If during expiration the two compartments empty synchronously, they can be functionally combined in one single compartment with the alveolar concentration F_A:

$$F_A = (V'_{A_1} \cdot F_{A_1} + V'_{A_2} \cdot F_{A_2})/(V'_{A_1} + V'_{A_2}) \qquad [A5.6]$$

The arterial CO_2 partial pressure is calculated in the same way taking the eq. A5.5 into consideration and, as is defined above, with $Q'_{c_1} = Q'_{c_2} = Q'/2$

$$P_a = (Q'_{c_1} \cdot P_{c_1} + Q'_{c_2} \cdot P_{c_2})/(Q'_{c_1} + Q'_{c_2}) = (P_{c_1} + P_{c_2})/2 \qquad [A5.7]$$

If it is possible to replace the two compartments with different V'_A/Q' by one single representative one, the following must apply for this compartment:

$$P_A = P_a \quad \text{or} \quad F_A = K \cdot P_a \qquad [A5.8]$$

where K is a factor to convert partial pressure to volume fraction of dry gas. Eq. A5.8 expresses that eqs. A5.6 and A5.7 must be identical, as they refer to one and the same compartment.

$$(V'_{A_1} \cdot F_{A_1} + V'_{A_2} \cdot F_{A_2})/(V'_{A_1} + V'_{A_2}) = (F_{A_1} + F_{A_2})/2 \qquad [A5.9]$$

The solution of eq. A5.9 is that V'_{A_1} equals V'_{A_2}, which produces a contradiction, for V'_{A_1} and V'_{A_2} per definition are not the same (eq. A5.2). It is therefore not possible to describe an inhomogenous system by the sole means of a homogenous system. This contradiction can be resolved by inserting a correction term CORR into eq. A5.6:

$$F_A = (V'_{A_1} \cdot F_{A_1} + V'_{A_2} \cdot F_{A_2})/(V'_{A_1} + V'_{A_2} + CORR) \qquad [A5.10]$$

Using eq. A5.10 instead of eq. A5.6 in eq. A5.8 the following applies:

$$(V'_{A_1} \cdot F_{A_1} + V'_{A_2} \cdot F_{A_2} + CORR) = (F_{A_1} + F_{A_2})/2 \qquad [A5.11]$$

From this CORR can be calculated as

$$CORR = 1/2 \cdot 1/F_A \cdot (F_{A_1} - F_{A_2}) \cdot (V'_{A_1} - V'_{A_2}) \qquad [A5.12]$$

The greater the inhomogeneity of the V_A/Q is, i.e., the greater the difference between V'_{A_1} and V'_{A_2} in relation to perfusion (Q remains constant here), the greater the difference between the alveolar concentration F_{A_1} and F_{A_2} and the greater the correction term CORR.

Thus CORR can be used to describe deviations of the gas exchange from the ideal, i.e., V'_A/Q' inhomogeneities. CORR represents nothing else than the ventilation of unperfused areas. Such ventilation which does not participate in the gas exchange is called *dead space ventilation.* The conclusion is that even if high dead space are measured, this does not prove the existence of compartments that are ventilated but not perfused (dead space). The same considerations apply for perfusion where inhomogeneity results in a *functional shunt.*

Determining F_{A_1} and F_{A_2} poses practical difficulties. First of all, the areas are not accessible and, secondly, they exist not as discrete compartments but as spatial distributions. These problems resulted in the concept of the "ideal" compartment and in the "ideal" gas equation (eq. 2.23).

Appendix 6

Viscosity of a Gas Mixture (Excerpt from Brunner 1983)

The viscosity of a gas refers to its property of offering resistance to the movement of adjacent layers (internal friction). The gases used here are called newtonian gases, as there is a linear relationship between shear stress and radial velocity gradients. The proportionality constant is the "dynamic viscosity" η. A change in the composition of a gas produces a change in its viscosity. Chapman (1960) claims that *no* closed mathematical expression can be given for the relationship between viscosity and gas composition. There are, however, numerous approximations for binary gas mixtures (Chapman 1960, Sutherland 1895, and Puluj 1879 in Schmitt 1909). In: Weast (ed) (1974–75) an empirical formula for mixtures with more than one compartment is given. Since, however, only the gases N_2, O_2, Ar, and CO_2 in small amounts are to be dealt with, the relationship between gas mixture and viscosity can be described using a linear approach. The viscosity of the dry gas mixture η_{dry} is then calculated as follows:

$$\eta_{dry} = \eta_{N_2} \cdot F_{N_2} + \eta_{O_2} \cdot F_{O_2} + \eta_{CO_2} \cdot F_{CO_2} + \eta_{Ar} \cdot F_{Ar} \qquad [A6.1]$$

F = the dry gas fraction of the gases.

Although the viscosity is nonlinearly dependent on temperature, a linear approximation suffices in the range between 20 and 40 degrees C:

$$\eta_{N_2} = 50.549 + 0.423 \cdot T \text{ (Chapman, Cowling 1960)} \qquad [A6.2]$$
$$\eta_{Ar} = 34.945 + 0.635 \cdot T \text{ (Chapman, Cowling 1960)} \qquad [A6.3]$$
$$\eta_{O_2} = 49.595 + 0.521 \cdot T \text{ (Kohlrausch 1968)} \qquad [A6.4]$$
$$\eta_{CO_2} = 11.197 + 0.461 \cdot T \text{ (Kohlrausch 1968)} \qquad [A6.5]$$

η in 10^2 Pa s, temperature T in degrees Kelvin (°K).

Using the above equations, the viscosities at 20 °C become 1.7449, 2.21, 2.0222, and 1.4627 Pa s for nitrogen, argon, oxygen, and carbon dioxide.

Example: the difference between the viscosity of oxygen/nitrogen mixture and oxygen/argon mixture at a constant O_2 content (21%) amounts to 0.258% per volume percent argon (see Chapter 3, Fig. 3.2).

Water vapor occurs only in small amounts. It is nevertheless taken into consideration here, as its viscosity is very different from that of other gases. A linear approach with the coefficients in eq. A6.6 describes an approximation of the dependency of viscosity on water vapor (η_{H_2O}):

$$\eta_{H_2O} = -5.55 + 0.35 \cdot T \quad \text{(Gieck 1981)} \qquad [A6.6]$$

η in 10^2 Pa s, temperature T in degrees Kelvin (°K).

The viscosity of wet gas is calculated from the equations A 6.2 to A 6.6 under the assumption of linearity to

$$\eta_{wet} = [rF \cdot P_{H_2O} \cdot \eta_{H_2O} + (P_B - rF \cdot P_{H_2O}) \cdot \eta_{dry}]/P_B \qquad [A\,6.7]$$

rF is the degree of saturation (between 0 and 1), P_{H_2O} is the water vapor pressure, and P_B the barometric pressure in torr. P_{H_2O} depends on the absolute temperature and can be calculated according to eq. 3.5.

Appendix 7

Determination and Dependencies of the Delay Time Between Flow-sensor and Mass Spectrometer (Excerpt From Brunner 1985c)

Mathematical model

Fig. A7.1 gives a block diagram of the mass spectrometer (see Lerou 1984). Gas is drawn from the measuring site into the mass spectrometer (MS) via the capillary. The driving force is the pressure difference between the measuring site (P_B) and the inlet system (P_V) of the mass spectrometer. After its entry, the flow of gas in the capillary is laminar up to the point where it enters the inlet system. At pressures of about 10^{-1} torr that exist here, molecular flow (Nudsen flow) is not to be expected. For incompressible gas, eq. A7.1 according to Hagen-Poiseuille applies:

$$dV/dt = \eta \cdot r^4/(8 \cdot \pi \cdot l) \cdot (P_B - P_V) \qquad [A7.1]$$

where dV/dt = volume flow, r = radius of capillary, l = length of capillary, η = dynamic viscosity of the gas.

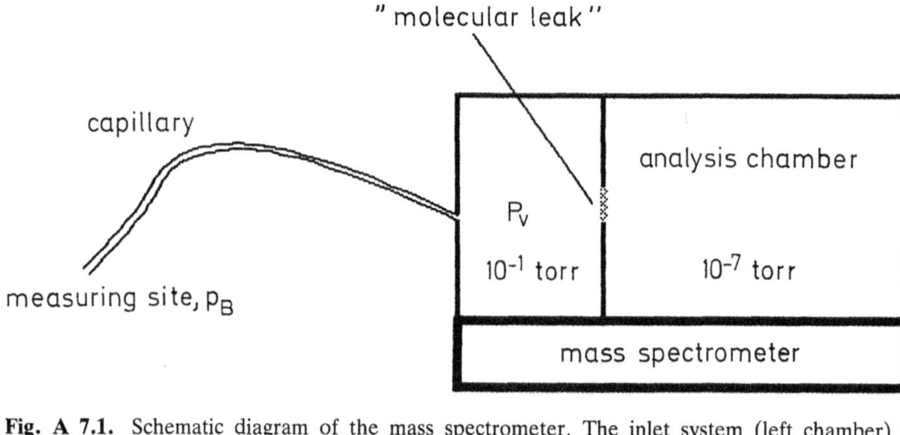

Fig. A 7.1. Schematic diagram of the mass spectrometer. The inlet system (left chamber) is evacuated by means of a rotation pump to $P_v = 10^{-1}$ torr. As a result of the pressure difference from that at the measuring site ($P_B - P_V$), the gas is transported into the inlet system. There is high-vacuum in the analysis chamber which is separated from the inlet system by a "molecular leak"

If a concentration step is produced at the measuring point, the velocity (V) of the gas front in the capillary when the viscosity, capillary geometry, and pressure difference are constant is calculated to be

$$v = dV/dt/(r^2 \cdot \pi) \qquad [A7.2]$$

Under the assumption that the transport time (T_v) of the gas front from the inlet system into the analyzing chamber and to the display on the instrument is constant, the total delay time (T_D) can be calculated as $T_D = l/v$ using eqs. A7.1 and A7.2:

$$T_D = \eta \cdot l^2 \cdot 8/r^2/(P_B - P_V) + T_v \qquad [A7.3]$$

For compressible gases, the equations no longer apply for the entire capillary, but only for the infinitesimal lengths (decompression effect, Boyle's and Mariotte's law: $p \cdot V = \text{const.}$). According to James and Martin (1952), this results in a correction factor of 4/3 under the assumption that $P_V \ll P_B$:

$$T_D = \eta \cdot l^2 \cdot 23/3/r^2/P_B + T_v \qquad [A7.4]$$

with T_d given in milliseconds.

Methods and Measurements

To measure the delay time T_D, a special 500 ml syringe is necessary (Fig. A7.2), with which a gas change can be produced at the tip of the capillary. This gas change takes place after a certain volume is expelled (50 ml here). The time from when this volume is pushed out to the moment when 50% of the concentration step appears on the MS is defined as T_D (Fig. A7.2). The volume is measured with the flow transducer, to which the delay time T_D will refer for later synchronization. To determine the dependency of T_D on the length of the capillary, one and the same capillary was shortened from 3 m to 1.4 m in 6 steps and measured 7 times each. On 4 capillaries of different lengths the T_D of a CO_2 interface (8%) was measured in N_2/O_2 and in Ar/O_2 (F_{O_2} constant 21%). The mean value for N_2/O_2 taken from 4 measurements per capillary was divided by the mean value for Ar/O_2 also taken from 4 measurements per capillary.

The viscosity of the gas mixture was calculated by linear superposition of the individual gas viscosities (Appendix 6: "Viscosity of a gas mixture"). The gas composition was measured with a Centronic MGA 200 mass spectrometer and the gas flow was measured with a Fleisch pneumotachograph No. 2. The analog data were stored on disk by means of an A/D converter (sample rate 60 Hz) and then analyzed, as shown in Fig. A7.2. The pneumotachograph and mass spectrometer were calibrated immediately prior to the measurements.

Results

With the calibration syringe, an abrupt gas change is produced. Although the steepness of the front generated is finite, it is nevertheless so quick that the mass spectrometer is not able to follow it. At the output of the mass spectrometer, therefore, its response time is measured (80 ms for 10 to 90% of the step response) and

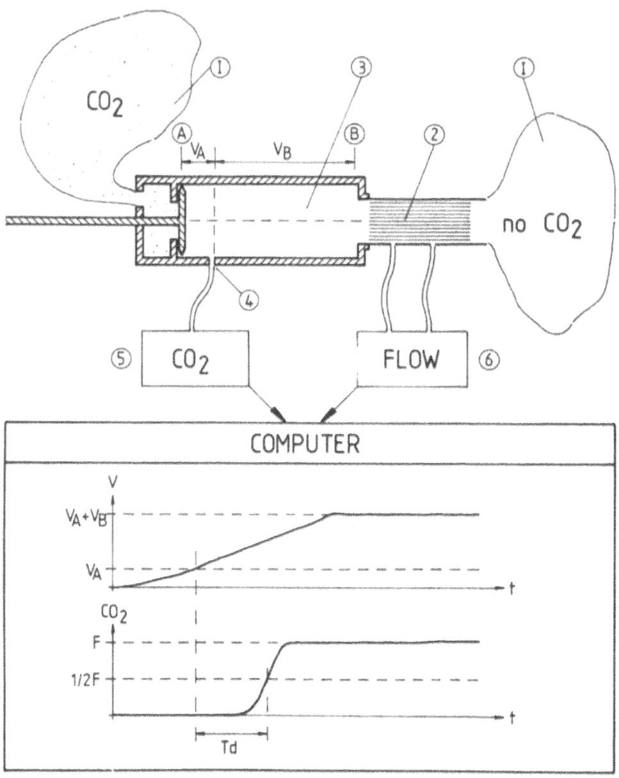

Fig. A 7.2. Set up of measuring equipment. Using the calibration syringe (3), a concentration step is generated at the tip of the capillary (4) along with a flow signal in the pneumotachograph (2) and a connected differential pressure transducer (6). A gas free of CO_2 and a mixture containing CO_2 are added in two plastic sacks (1) on both sides of the piston. If the piston is moved from A to B, the CO_2 front is generated after V_A is expelled. The time to when 50% of the end-value is displayed on the CO_2 sensor is called delay time (T_d)

not the steepness of the gas front generated with the syringe. This gas front can be considered ideal for the measurements. Table A7 gives the mean $T_D \pm$ SD of the 7 measurements made with each of the 6 capillary lengths.

The longer the capillary the greater the delay time. Using the least squares method, the values given in Table A7 can be represented as

$$T_D = 58.57 \cdot l^2 + 61 \; (r^2 = 0.998) \qquad [A7.5]$$

(with r^2 being the coefficient of correlation). The theoretical value for the slope, as calculated using eq. A7.4, is 56.2 ms/m² (with $\eta = 1.82\ 10^{-5}$ N s/m², $P_B = 95.76$ kPa or 720 torr and an inner radius of the polyethylene capillary of r = 0.19 mm).

The mean T_D of the 4 capillaries which are measured with the 2 gases having different viscosities was 490 ± 73.4 ms (CO_2 front of 8 vol% in N_2/O_2 mixture); in

Table A 7. Dependence of delay time on length of capillary

length	3.0 m	2.5 m	2.0 m	1.8 m	1.6 m	1.4m
T_D [ms]	585.3	433.3	309.2	245.8	207.4	174.0
SD [ms]	9.6	—	8.8	7.7	8.8	8.8

Ar/O_2 mixture it was 19.7 ± 2.4% longer. Thus the increase in the viscosity of 19.23% resulted in an increase in the T_D of 19.7 ± 2.4%. Also considered in calculating the percent values was a constant term T_v of 61 ms.

Discussion

The purpose of the calibration syringe described is to produce a gas change at the measuring point of the mass spectrometer, as rapid as possible and at a defined instant in time. The gas front generated is steep enough for our purposes, as the measured rise time corresponds to the fastest rise time of the mass spectrometer.

The uncertainty in determining the delay time stems from the sampling rate, the way the MS works, and the phase shift in the pneumotachograph. Every 20 ms, the MS puts out a new measured value per channel. This results in a discretization error of ± 10 ms. According to Jackson (1979), the phase shift in the pneumotachograph increases with increasing frequency and in air amounts to less than 20 degrees at 20 Hz. This is less than 3 ms and is therefore negligible. The measurements were reproducible to within better than 9.6 ms. This suggests that the accuracy is limited by the sampling period of the mass spectrometer.

The constant term T_v in the eqs. A7.4 and A7.5 (61 ms) is composed of the transport time from the inlet vacuum into the high vacuum of the MS together with the rise time from 0 to 50% of the end value. Provided T_v is known, it is possible to calculate T_D from the specifications of the capillary using eq. A7.4. The problem is, however, that the capillary radius r is only roughly specified (0.19 ± 0.04 mm). And as in eq. A7.4 r is squared, the calculated delay time can deviate from the actual value by more than 100%. The fact that the experimental and the theoretical value agreed in this study (58.57 and 56.2 ms/m^2, respectively), is probably just coincidence. Since r is so inexactly determined and is squared in eq. A7.4, T_D should not be calculated but only measured.

The results of the measurements qualitatively agree with the simple mathematical model (Fig. A7.1). T_D is directly proportional to the viscosity of the gas mixture. In clinical practice, F_{IO_2} varies between 20 and 100% for therapeutic reasons. During the N_2 washout with Ar/O_2, the viscosity can change as much as 20%. This has an immediate effect on the measurements of PIE, alv.eff.-CO_2, V'_{CO_2}, and APV (in the N_2 washout), with errors of up to 70% (Brunner 1985c). For this reason, T_D should be adapted breath-by-breath to the actual gas viscosity. We use the following procedure, provided $T_D(o)$ for the viscosity η_o is measured with the method described above.

1. Determine the end-expiratory gas composition F_{N_2et}, F_{O_2et}, and F_{CO_2et} of the breath n,
2. calculate the viscosity η_n of this gas mixture using eq. A6.7,
3. calculate $T_D(n)$ for the current breath n according to eq. A7.6

$$T_D(n) = (T_D(o) - 61) \cdot \eta_n/\eta_o + 61 \qquad [A7.6]$$

$T_D(o)$ in milliseconds is the delay time *measured* with the gas mixture η_o.

All of the instantaneous values (from the A/D converter) of the nth breath are now delayed by the whole number portion of the quotient $T_D(n)/16.66$, except those from the MS. Accordingly, the smallest time unit that can be corrected is 16.66 ms (sample rate 60 Hz). The procedure corresponds to method IV in Brunner (1985c) and provides the accuracy of PIE and APV determinations as is described in detail in Chapter 3.3.1.

Appendix 8

Mathematical Formulas to Calculate Lung Function Indices

This chapter is intended as reference to anyone who wants to compute the lung function indices we used in this book. For simplicity, the references have been left out in this chapter but are given in the text.

For users of the Basle-Software: The abbreviations are chosen to allow easy reference to the FORTRAN-IV program variables. This is the reason why, for example, barometric pressure is called TEMPBY (for temperature, body) rather than the conventional T_B.

All signals are read from the A/D converter and stored in arrays. The notion x(n) therefore means the nth sample of x as well as the nth element in the array x. Since A/D conversion is done at a constant sample rate, n can be converted to time by dividing it by the sample rate (here 60 Hz).

Definitions

#	"number"
n	integer number signifying the nth sample from the A to D converter
i	integer number signifying the ith sample in a breath-phase
K	number of digital samples acquired during a continuous period of time
$\Sigma x(n)$	n = 1..K Sum of all samples of x starting at # 1 and ending with # K
I,E	number of digital samples acquired during inspiration and expiration respectively, also used as subscript for inspiration and expiration
ADU(N,n)	n^{th} Analog-Digital-Unit, integer number between 0 and 4095 of analog input channel # N
T_D	Delay time (see appendix) in ms
T	current gas temperature in °K
TEMPBY	body temperature in °C
TEMPAM	room temperature °C
P_B	barometric pressure in Torr
rF	relative humidity in %
i = 1..I	means "over entire inspiration"
i = 1..E	means "over entire expiration"
i = 1..I..E	means "over entire breath"

Analog input signals and their associated channel on the A/D board

N = 1	airway flow · sample interval (volume increments), V in ml/sec · 16.666 ms
N = 2	N_2 concentration F_{N_2} in fraction (0..1)
N = 3	CO_2 concentration F_{CO_2} in fraction
N = 4	O_2 concentration F_{O_2} in fraction
N = 5	airway pressure P_{awo} in cmH_2O
N = 6	esophageal pressure P_{es} in cmH_2O

Conversion factors STPD:BTPS

Assumptions		
Inspiration	$T = 273 + TEMPAM + 5$	[°K]
	Pressure is P_B	[Torr]
	rF = 60%	
Expiration	$T = 273 + TEMPBY - 5$	[°K]
	Pressure is P_B	[Torr]
	rF = 80%	

V_{STPD}	$= V_{BTPS} \cdot 273/760 \cdot (P_B - P_{H_2O})/T$	[ml]
with P_{H_2O}	$= 13.2 - 0.61 \cdot (T - 273) + 0.04 \cdot (T - 273)^2$	[Torr]

If the relative humidity (rF) is less than 100%, P_{H_2O} must be multiplied by that percentage and divided by 100%.

1. Calibration procedure

Assumption: linear relationship of the form:
$x(N,n) = a \cdot ADU(N,n) + b$,
where N is the channel number (see above) and n is the nth sample. The constants a and b are calculated using two points in the x vs. ADU diagram.

For users of the Basle-Software: The x-values are found in the file HEAD.DAT where they were entered using the program SPEC. The ADU-values are calculated during the calibration session when the user marks two portions of data for each channel on the screen. Although the procedure is essentially the same for all channels, the flow calibration differs slightly from the others.

To obtain volume in ml from an ADU, the constants BASELN, KVOLIN and KVOLEX have to be determined as follows:

The calibration volume V_{cal} is applied by a calibrated syringe and measured simultaneously. The user specifies periods of no-flow (K_O), inspiratory flow (K_I) and expiratory flow (K_E). The ADU's measured during those periods are used to derive BASELN, KVOLIN and KVOLEX.

BASELN	$= 1/K_O \cdot \Sigma ADU(1,n)$	$n = 1 \ldots K_O$
KVOLIN	$= V_{cal}/\Sigma[ADU(1,n) - BASELN]$	$n = 1 \ldots K_I$
KVOLEX	$= V_{cal}/\Sigma[ADU(1,n) - BASELN]$	$n = 1 \ldots K_E$

In the current program version the mean value of KVOLIN and KVOLEX is used for both, inspiration and expiration: KVOL = (KVOLIN + KVOLEX)/2.

To obtain fractions of nitrogen from an ADU, the constants K_{N_2A} and K_{N_2B} have to be determined as follows:

Two concentrations of nitrogen F_1 and F_2 are measured. The user specifies the data sections K_1 and K_2 which correspond to F_1 and F_2. The ADU's measured during those periods are averaged and used to derive K_{N_2A} and K_{N_2B}.

Caution is required using calibration gases which contain CO_2 (see text, cracking pattern). If CO_2 is present, the reading on the mass spectrometer should be entered rather than the actual nitrogen concentration in the calibration gas.

K_{N_2A}	$= (F_1 - F_2)/[\Sigma ADU(2,n_1)/K_1 - \Sigma ADU(2,n_2)/K_2]$	$n_1 = 1 \ldots K_1$, $n_2 = 1 \ldots K_2$
K_{N_2B}	$= F_1 - K_{N_2A} \cdot \Sigma ADU(2,n_1)/K_1$	$n_1 = 1 \ldots K_1$

To obtain fractions of CO_2 from an ADU, the constants K_{CO_2A} and K_{CO_2B} have to be determined as follows:

Two concentrations of carbon dioxide F_1 and F_2 are measured. The user specifies the data sections K_1 and K_2 which correspond to F_1 and F_2. The ADU's measured during those periods are averaged and used to derive K_{CO_2A} and K_{CO_2B}.

K_{CO_2A}	$= (F_1 - F_2)/[\Sigma ADU(3,n_1)/K_1 - \Sigma ADU(3,n_2)/K_2]$	$n_1 = 1 \ldots K_1$, $n_2 = 1 \ldots K_2$
K_{CO_2B}	$= F_1 - K_{CO_2A} \cdot \Sigma ADU(3,n_1)/K_1$	$n_1 = 1 \ldots K_1$

To obtain fractions of oxygen from an ADU, the constants K_{O_2A} and K_{O_2B} have to be determined as follows:

Two concentrations of oxygen F_1 and F_2 are measured. The user specifies the data sections K_1 and K_2 which correspond to F_1 and F_2. The ADU's measured during those periods are averaged and used to derive K_{O_2A} and K_{O_2B}.

K_{O_2A}	$= (F_1 - F_2)/[\Sigma ADU(4,n_1)/K_1 - \Sigma ADU(4,n_2)/K_2]$	$n_1 = 1 \ldots K_1$, $n_2 = 1 \ldots K_2$
K_{O_2B}	$= F_1 - K_{O_2A} \cdot \Sigma ADU(4,n_1)/K_1$	$n_1 = 1 \ldots K_1$

To obtain airway pressure in cmH_2O from an ADU, the constants K_{PRESA} and K_{PRESB} have to be determined as follows:

Two levels of pressure P_1 and P_2 are measured. The user specifies the data sections K_1 and K_2 which correspond to P_1 and P_2. The ADU's measured during those periods are averaged and used to derive K_{PRESA} and K_{PRESB}.

$K_{PRESA} = (P_1 - P_2)/[\Sigma ADU(5,n_1)/K_1 - \Sigma ADU(5,n_2)/K_2]$ $n_1 = 1 \ldots K_1$, $n_2 = 1 \ldots K_2$

$K_{PRESB} = P_1 - K_{PRESA} \cdot \Sigma ADU(5,n_1)/K_1$ $n_1 = 1 \ldots K_1$

To obtain esophageal pressure in cmH_2O from an ADU, the constants K_{ARA} and K_{ARB} have to be determined as follows:

Two levels of pressure P_1 and P_2 are measured. The user specifies the data sections K_1 and K_2 which correspond to P_1 and P_2. The ADU's measured during those periods are averaged and used to derive K_{ARA} and K_{ARB}.

$K_{ARA} = (P_1 - P_2)/[\Sigma ADU(6,n_1)/K_1 - \Sigma ADU(6,n_2)/K_2]$ $n_1 = 1 \ldots K_1$, $n_2 = 1 \ldots K_2$

$K_{ARB} = P_1 - K_{ARA} \cdot \Sigma ADU(6,n_1)/K_1$ $n_1 = 1 \ldots K_1$

This completes the determination of the calibration constants.

2. Calculation of viscosity of a gas mixture

Viscosity of dry gas is

$$\eta_{dry} = F_{N_2} \cdot (50.549 + 0.423 \cdot T) + F_{O_2} \cdot (49.595 + 0.521 \cdot T) + F_{CO_2} \cdot (11.197 + 0.461 \cdot T) + F_{Ar} \cdot (34.945 + 0.635 \cdot T)$$

η in 10^2 Pa s, T in °K.

Viscosity of wet gas with relative humidity rF is:

$$\eta_{wet} = [rF \cdot P_{H_2O} \cdot (-5.55 + 0.35 \cdot T) + (P_B - rF \cdot P_{H_2O}) \cdot \eta_{dry}]/P_B$$

3. Scaling and correction of ADU's during analysis (subroutine INVAL)

Assumption: The pneumotachograph was calibrated with air. The delay-time was measured with *exhaled air*. In both cases the gas viscosity is close to the viscosity of room air.

Viscosity of room air: VISCOS = $\eta_{wet}(air)$
Viscosity of current sample n: VISC = $\eta_{wet}(F_{CO_2}(n), F_{N_2}(n), F_{O_2}(n), F_{Ar}(n))$
Measured Delay time: T_D
Current Delay time: $T_D = (T_D - 61) \cdot VISC/VISCOS + 61$
Number of samples to be delayed: DEL = $T_D/16.666$

The nth samples for gas composition are given by:
$F_{CO_2}(n) = K_{CO_2A} \cdot ADU(3,n + DEL) + K_{CO_2B}$
$F_{N_2}(n) = K_{N_2A} \cdot ADU(2,n + DEL) + K_{N_2B} - 0.1 \cdot F_{CO_2}(n)$
$F_{O_2}(n) = K_{O_2A} \cdot ADU(4,n + DEL) + K_{O_2B}$

The nth volume element is:
$V(n) = (ADU(1,n) - BASELN) \cdot KVOL \cdot VISCOS/VISC$

The nth pressure samples are:
$P_{awo}(n) = K_{PRESA} \cdot ADU(5,n) + K_{PRESB}$
$P_{oes}(n) = K_{ARA} \cdot ADU(6,n) + K_{ARB}$

4. Derivation of basic values

Inspiratory volume $V_I = \Sigma V(i)$	[ml], i = 1 ... I
Expiratory volume $V_E = \Sigma V(i)$	[ml], i = 1 ... E
Tidal volume $V_T = (-V_I + V_E)/2$	[ml]
Inspiratory time $T_I = I \cdot 16.66$	[ms]
Expiratory time $T_E = E \cdot 16.66$	[ms]
Respiratory Rate $RR = 60/(T_I + T_E)$	[l/min]
Minute Volume $MV = V_T \cdot RR/1000$	[l/min]
Mean pressure P_{awo}-mean $= \Sigma P_{awo}(i)/(T_I + T_E)$	[cmH_2O], i = 1 ... I ... E
End-inspiratory pressure P_{awo}-EI $= P_{awo}(I - 1)$	[cmH_2O]
End-expiratory pressure P_{awo}-EE $= 1/5 \cdot \Sigma P_{awo}(e)$	[cmH_2O], e = E − 4 ... E

An alternative method for P_{awo}-EE is given in section 6 about lung mechanics.

CO_2 elimination per breath $V_{CO_2} = \Sigma[V(i) \cdot F_{CO_2}(i)]$	[ml], i = 1 ... I ... E
Oxygen uptake per breath $V_{O_2} = \Sigma[V(i) \cdot F_{O_2}(i)]$	[ml], i = 1 ... I ... E
N_2 washed out per breath: $V_{N_2} = \Sigma[V(i) \cdot F_{N_2}(i)]$	[ml], i = 1 ... I ... E

Caution: Care must be taken in calculating V_{O_2} because of the respiratory quotient ($V_I \neq V_E$). Example: at F_{IO_2} of 100 vol%, V'_{CO_2} of 200 ml/min and R of 0.85 the difference $V_I - V_E$ is 30 ml or 0.3% for V_T of 1000 ml. The above equations assume, that flow can be measured with very high precision, i.e., better than 1%. Since this is not achieveable with the present set up, formulas are used which are derived in the following.

A weighted average of inspiratory and expiratory concentrations is calculated first (per breath):

F_{IO_2}	$= 1/V_I \cdot \Sigma F_{O_2}(i) \cdot V(i)$	[vol%], i = 1 ... I;
F_{EO_2}	$= 1/V_E \cdot \Sigma F_{O_2}(e) \cdot V(e)$	[vol%], e = 1 ... E;
F_{ICO_2}	$= 1/V_I \cdot \Sigma F_{CO_2}(i) \cdot V(i)$	[vol%], i = 1 ... I;
F_{ECO_2}	$= 1/V_E \cdot \Sigma F_{CO_2}(e) \cdot V(e)$	[vol%], e = 1 ... E.

It is assumed that the net exchange of nitrogen is zero ($V_{IN_2} = V_{EN_2}$, steady state). This leads to the equations for V'_{O_2} and V'_{CO_2}:

$$V_{O_2} = \frac{(V_E \cdot F_{IO_2} - V_E \cdot F_{EO_2} - V_E \cdot F_{ECO_2} \cdot F_{IO_2} + V_E \cdot F_{EO_2} \cdot F_{ICO_2})}{(1 - F_{IO_2} - F_{ICO_2})} \text{ [ml]}$$

$$V_{CO_2} = \frac{(V_E \cdot F_{ECO_2} - V_E \cdot F_{ICO_2} - V_E \cdot F_{ECO_2} \cdot F_{IO_2} + V_E \cdot F_{EO_2} \cdot F_{ICO_2})}{(1 - F_{IO_2} - F_{ICO_2})} \text{ [ml]}$$

During the nitrogen washout, the amount of N_2 washed out per breath (V_{N_2}) is measured. The assumption of no net transfer of nitrogen is evidently not valid. Since the rebreathing of nitrogen has to be kept to a minimum, the following equations are accurate enough to calculate V_{N_2} provided the FRC does not change during the washout:

$F_{IN_2} = 1/V_I \cdot \Sigma F_{N_2}(i) \cdot V(i)$ [vol%], i = 1 ... I;
$F_{EN_2} = 1/V_E \cdot \Sigma F_{N_2}(e) \cdot V(e)$ [vol%], e = 1 ... E;

$V_{N_2} = (V_E \cdot F_{EN_2} - V_E \cdot F_{IN_2})$ [ml]

This completes the description of the V_{O_2}, V_{CO_2} and V_{N_2} calculations.

5. Lung volume and washout efficiency

A nitrogen washout is analyzed until the end tidal concentration falls below 1 vol% (otherwise noted in the text). The number of breaths required for this is referred to as M.

Net N_2-transfer in breath m ($V_{N_2}(m)$) is calculated as described above and the Accessible Pulmonary Volume is calculated thereof:

$APV = \Sigma V_{N_2}(m)/(F_{N_2}(O) - F_{N_2}(M))$ [ml], m = 1 ... M

Washout efficiency:
$WO.eff. = TO_{ideal}/TO_{real} \cdot 100$ [%]

Moment analysis, modified according to text ($F_{N_2alv}(0) = F_{N_2}{}^{o}$, $TO(0) = 0$, V_{ext} see Fig. 2.7).

$F_{N_2alv}(m) = V_{N_2}(m)/[V_E(m) - V_{dS}(m)]$ m = 0, 1, 2, 3 ...
$Y(m) = F_{N_2alv}(m)/F_{N_2}{}^{o}$ m = 0, 1, 2, 3 ...

$TO(m) = \Sigma[V_E(k) - V_{dS}(k) - V_{ext}(k)]/[APV + V_{ext}(k)]$ k = 1 ... m

$\mu_r = \Sigma Y(m) \cdot [TO(m+1) - TO(m)] \cdot TO(m)^r$ m = 1 ... M,
r = 0, 1, 2 ...th moment

Normalization of moments is done by dividing the rth moment by μ_O.

6. Lung mechanics by least squares fitting

Basic equation:
$P_{awo}(t) = V(t)/C_{tot} + dV(t)/dt \cdot R_{tot} + P_{awo}\text{-EE}$ [cmH_2O]

Abbreviations and Definitions:
$Vol(i) = \Sigma[-V(k)]$ [ml], k = 1 ... i
$L = I + E$ number of samples
$\tau = 16.666$ [ms]

$a = \Sigma[\ V(i)/\tau \cdot V(i)/\tau]$ i = 1 ... I ... E
$b = \Sigma[-Vol(i) \cdot V(i)/\tau]$ i = 1 ... I ... E
$c = \Sigma - V(i)/\tau$ i = 1 ... I ... E
$d = \Sigma[-V(i)/\tau \cdot P_{awo}(i)]$ i = 1 ... I ... E

$e = \Sigma[Vol(i) \cdot P_{awo}(i)]$ $\quad i = 1 \ldots I \ldots E$
$f = \Sigma P_{awo}(i)$ $\quad i = 1 \ldots I \ldots E$
$g = \Sigma[Vol(i) \cdot Vol(i)]$ $\quad i = 1 \ldots I \ldots E$
$h = \Sigma Vol(i)$ $\quad i = 1 \ldots I \ldots E$

$$C_{tot} = \frac{2 \cdot b \cdot c \cdot h - c \cdot c \cdot g - a \cdot h \cdot h + L \cdot a \cdot g - L \cdot b \cdot b}{c \cdot d \cdot h - c \cdot c \cdot e + L \cdot a \cdot e - L \cdot b \cdot d - a \cdot f \cdot h + b \cdot c \cdot f} \text{ [ml/cmH}_2\text{O]}$$

The same formula is used to calculate lung compliance C_L. In this case, $P_{awo}(i)$ must be substituted by the difference $P_{awo}(i) - P_{es}(i)$.

$$P_{awo}\text{-EE} = \frac{e \cdot a - b \cdot d - (g \cdot a - b \cdot b)/C_{tot}}{h \cdot a - b \cdot c} \text{ [cmH}_2\text{O]}$$

$$R_{tot} = \frac{d - P_{awo}\text{-EE} \cdot c - b/C_{tot}}{a} \cdot 1000 \text{ [cmH}_2\text{O/(l/sec)]}$$

7. Calculations using the CO_2-diagram

PIE can be determined according to the formula in the text or with the simple algorithm as follows.

Definition of sample #i_1 and sample #i_2: At sample #i_1 the CO_2-concentration is assumed to reach $0.5 \cdot F_{CO_2\text{-et}}$. The volume exhaled until $0.5 \cdot F_{CO_2\text{-et}}$ is reached can be calculated as $PIE_{approx} = \Sigma V(i), i = 1 \ldots i_1$. At sample #$i_2$ the volume $2 \cdot PIE_{approx}$ is expired which is expressed by $2 \cdot PIE_{approx} = \Sigma V(i), i = 1 \ldots i_2$. E.g. focusing on phase 2 (first step as described in the text) yields i_2.

$V_{CO_2}(i_2) = \Sigma[V(i) \cdot F_{CO_2}(i)]$ $\quad$ [ml], $i = 1 \ldots i_2$
$PIE = 2 \cdot PIE_{approx} - V_{CO_2}(i_2)/F_{CO_2}(i_2)$ $\quad$ [ml]
$F_{CO_2 alv.} = V_{CO_2}/(V_E - PIE) \cdot 100$ $\quad$ [vol%]
$alv.eff_{CO_2}. = F_{CO_2 alv.} \cdot (P_B - P_{H_2O})/P_{aCO_2} \cdot 100$ $\quad$ [%]

The calculation of the slope of phase III requires foccusing on the alveolar plateau e.g., the determination of the second and third quarters of phase 3 and the corresponding boundaries in terms of sample numbers $i_{1/4}$, $i_{2/4}$ and $i_{3/4}$.

Definitions:
$Vol(i) = \Sigma V(k)$ $\quad$ [ml], $k = 1 \ldots i$
$a = \Sigma Vol(i)/(i_{2/4} - i_{1/4})$ $\quad i = i_{1/4} \ldots i_{2/4}$
$b = \Sigma Vol(i)/(i_{3/4} - i_{2/4})$ $\quad i = i_{2/4} \ldots i_{3/4}$
$c = \Sigma F_{CO_2}(i)/(i_{2/4} - i_{1/4})$ $\quad i = i_{1/4} \ldots i_{2/4}$
$d = \Sigma F_{CO_2}(i)/(i_{3/4} - i_{2/4})$ $\quad i = i_{2/4} \ldots i_{3/4}$

$slope_{CO_2} = (d - c)/(b - a) \cdot O^5$ $\quad$ [vol%/l]

The same formulas apply for the N_2 diagram.

8. Calculations using oxygen measurement in gas and blood

These formulas are the classical expressions derived by Riley and co-workers. The algorithms published by Kelman are used.
Alveolar Gas equations (in steady state):

$P_{ACO_2} = P_{aCO_2}$ [Torr]

$P_{ACO_2} = F_{IO_2} \cdot (P_B - P_{H_2O}) - P_{ACO_2} \cdot [1 + F_{IO_2} \cdot (1 - R)]/R$ [Torr]

To calculate the oxygen saturation corresponding to a particular P_{O_2}, the measured P_{O_2} has first to be converted into a virtual P_{O_2virt} in order to be applicable to a standard dissociation curve. Temperature T in °C.

$$P_{O_2virt} = P_{O_2} \cdot 10^{[0.024 \cdot (37-T)+0.4 \cdot (pH-7.4)+0.06 \cdot (\log 40-\log P_{CO_2})]}$$

An additional shift of the dissociation curve is caused by increased red cell concentration of 2,3-DPG. To compensate for this, the difference between the P_{50} of the patient and the P_{50} of the standard dissociation curve (26.8 Torr) must be known (DP50). The compensation is then easily done using the equation:

$$P_{O_2virt} = P_{O_2} \cdot 26.8/(26.8 + DP50)$$

Abbreviations and Definitions:

$P = P_{O_2virt}$
A = − 8′532.2289
B = 2′121.401
C = − 67.073989
D = 935′960.87
E = − 31′346.258
F = 2′396.1674
G = −67.104406

$$S_{O_2} = \frac{A \cdot P + B \cdot P^2 + C \cdot P^3 + P^4}{D + E \cdot P + F \cdot P^2 + G \cdot P^3 + P^4} \quad [\%]$$

If the P_{O_2} falls bellow 10 Torr, the above equation should be replaced by the following more accurate expression:

$$S_{O_2} = 0.003683 \cdot P + 0.000584 \cdot P^2 \quad [\%]$$

Content of oxygen:

$C_{O_2} = 1.34 \cdot Hb \cdot S_{O_2}/100 + \text{solubility} \cdot P_{O_2}$ [ml_{O_2}/ml_{blood}]

where solubility (ml_{O_2}/ml_{blood}/Torr) of oxygen is temperature (in °C) dependant:

$$\text{solubility} = 0.0059519 - 0.0001266 \cdot T + 0.0000013 \cdot T^2$$

Cardiac output (C. O.) and fraction of venous admixture (Q'_S/Q'_T) are calculated by measuring mixed venous saturation ($S_{\bar{v}O_2}$), arterial partial pressures of O_2 (P_{aO_2}) and CO_2 (P_{aCO_2}), blood pH (pH_a) and uptake of oxygen (V'_{O_2}). The oxygen contents of arterial (C_{aO_2}), mixed venous ($C_{\bar{v}O_2}$) and ideal arterial (C_{AO_2}) blood are then calculated using the above equations.

$C.O. = V'_{O_2}/(C_{aO_2} - C_{\bar{v}O_2})/1000$ [l/min]

$Q'_S/Q'_T = (C_{AO_2} - C_{aO_2})/(C_{AO_2} - C_{\bar{v}O_2})$

9. Correlations with anthropometric data

Body Surface (b. s. in m^2) is calculated using body height (b. h. in cm) and body weight (b. w. in kg):

$$\text{b.s.} = e^{[\ln(0.0235) + 0.422 \cdot \ln(\text{b.h.}) + 0.51456 \cdot \ln(\text{b.w.})]} \quad [m^2]$$

Expected FRC in the upright position (sitting) is correlated to age (A in years) and b. h. (in cm):

$$\text{expected FRC}_{sit} = 23.4 \cdot \text{b.h.} + 9 \cdot A - 1090 \quad \text{for men [ml]}$$
$$\text{expected FRC}_{sit} = 22.4 \cdot \text{b.h.} + A - 1000 \quad \text{for women [ml]}$$

Expected FRC in supine position is:

$$\text{expected FRC}_{sup} = \text{expected FRC}_{sit} \cdot 0.7$$

Expected FRC in supine position of anesthetized patients during controlled ventilation:

$$\text{expected FRC}_{sup\text{-}anest} = \text{expected FRC}_{sup} \cdot 0.85$$

V_{ds} is related to body height according to
$\text{expected } V_{ds} = 7.59 \cdot 10^{-4} \cdot \text{b.h.}^{2.36}$ [ml]

Appendix 9: Table 1: Standard values after cardiac surgery

Antropometric and general data													Haemodynamic Data mean blood pressures				
Patient	Sex	Indication	Procedure	Age	Weight	Height	Body Surface	LVEDP	Tubus ∅	Invest	Temp		Heart-R	art. BP	RAP	LAP	PAP
				[y]	[kg]	[cm]	[m²]	[mmHg]	[mm]	[hours]	[°C]		[/min]	[mmHg]	[mmHg]	[mmHg]	[mmHg]
1	m	AP+IMPI	ACB−5	53	90	173	2.10	18	9	3.5	35.9		90	82	7	8	19
2	m	AP	ACB−3	46	71	173	1.86	20	9	5.0	38.0		100	80	9	10	–
3	m	AP+ANEU	ACB−2+Aneu	61	74	169	1.88	22	9	7.0	39.3		100	75	10	13	26
4	m	AP	ACB−2	59	86	178	2.08	17	9	2.5	35.8		90	86	3	5	22
5	m	AP+pMIF	ACB−1+Symp	38	104	179	2.29	18	9	5.5	38.7		96	95	12	12	18
6	m	MI	ACB−1+MVR	68	83	186	2.08	15	9	2.3	35.8		102	77	10	12	32
7	m	AP+pMIF	ACB−2	59	111	198	2.48	12	9	1.3	34.5		79	84	7	15	17
8	m	AP	ACB−2	61	69	164	1.80	17	9	2.5	34.6		102	89	8	8	17
9	m	AP+pMIF	ACB−3	73	82	170	1.99	10	9	1.8	34.0		88	88	5	6	21
10	m	AP+pMIF	ACB−4	53	71	169	1.84	8	8	2.0	34.4		89	104	7	6	22
11	m	AP+pMIF	ACB−5	61	89	173	2.09	4	9	3.8	36.6		112	89	8	9	20
12	f	MS+AI	MVR+AVR	46	63	152	1.66	17	7	1.3	35.0		87	71	9	9	20
13	m	AP	ACB−3	61	87	186	2.13	18	9	1.8	35.8		90	90	5	7	23
14	m	AP+pMIF	ACB−5	65	85	172	2.04	9	8	1.5	35.5		84	111	6	7	18
15	m	AP+IMP	ACB−2	48	81	175	2.00	14	8	1.5	35.0		90	109	7	7	18
16	m	AP+pMIF	ACB−2	50	74	174	1.90	7	8	6.5	36.7		103	95	4	4	17
17	f	AP+IMP	ACB−1	67	78	163	1.90	16	7	1.0	35.6		74	52	7	6	17
18	m	AI	AVR	67	74	172	1.90	11	9	4.8	39.1		79	80	9	10	21
19	m	AIS	AVR	60	69	173	1.83	17	9	2.8	36.2		99	100	3	11	19
20	m	AP	ACB−3	70	74	164	1.86	17	9	7.8	34.5		90	79	12	15	24
21	f	AP+pMIF	ACB−3	63	86	164	2.01	17	8	5.5	35.0		90	72	11	10	22
22	f	MS	MVR	56	63	165	1.71	20	8	2.5	35.7		85	79	11	13	23
23	m	AP+pMIF	ACB−6	51	78	163	1.90	11	8	2.0	35.0		101	76	7	10	18
24	m	MI+AI	MVR+AVR	65	79	175	1.97	7	8	3.5	35.0		100	80	8	15	25
25	m	MSI	MVR+AVR	58	78	178	1.97	12	8	5.3	37.0		98	83	8	15	21
26	m	AP+pMIF	ACB−4	55	70	172	1.84	11	8	4.0	34.7		90	87	8	6	17
mean				58	80	172	1.97	14		3.4	35.9		93	85	8	10	21
SD				8	11	9	0.17	5		1.9	1.5		9	13	2	3	4
CV				14.4%	14.0%	5.2%	8.8%	33.5%		57.1%	4.1%		9.5%	14.8%	32.0%	35.2%	17.3%

Appendix 9: Table 1 (continued)

Ventilatory setting ($F_{IO_2} = 0.4$; PEEP = 0; EIP = 0)				Mechanical parameters							
RR [/min]	V_T [ml]	V'_I [ml/sec]	I:E [%]	MV [l/min]	$P_{awo\text{-}max}$ [cmH_2O]	$P_{awo\text{-}mean}$ [cmH_2O]	C_{tot} [ml/cmH_2O]	C_L	$R_{aw}+R_L$ [cmH_2O*s/l]	R_{tot}	R_{TUBUS} (estimate)
7.8	1394	562	32	10.9	22.0	5.6	77	–	–	8.0	2.4
7.6	1221	612	25	9.3	18.6	4.5	74	–	–	4.7	2.4
7.6	1162	617	24	8.8	16.9	4.6	85	156	4.0	5.5	2.4
6.5	1078	233	50	7.0	19.5	6.7	61	111	6.1	7.6	2.4
7.5	1223	322	47	9.2	26.8	9.7	50	–	–	7.6	2.4
7.8	1034	493	27	8.1	18.7	4.6	60	82	7.7	7.8	2.4
6.1	970	691	14	5.9	16.5	2.4	76	219	4.7	6.7	2.4
5.8	1214	227	52	7.0	17.9	6.2	73	106	6.2	7.1	2.4
6.4	917	310	32	5.9	19.1	4.0	52	80	3.9	5.4	2.4
6.2	859	285	31	5.3	13.3	3.0	70	124	4.2	5.3	3.1
8.9	1003	306	49	8.9	15.8	6.0	70	128	5.1	6.6	2.4
7.5	858	394	27	6.4	25.1	5.3	40	69	12.4	13.6	3.9
5.8	1130	337	32	6.6	20.8	5.3	62	91	10.7	11.2	2.4
6.9	1206	365	38	8.3	21.6	5.6	67	123	8.5	10.9	3.1
6.2	1186	361	34	7.4	21.6	5.4	64	98	10.4	11.4	3.1
6.9	1179	357	38	8.1	16.2	4.5	86	133	5.8	7.2	3.1
6.9	862	451	22	5.9	19.1	3.4	54	97	11.3	10.9	3.9
7.5	1208	429	35	9.1	17.8	4.7	74	183	5.5	5.8	2.4
7.6	1038	430	31	7.9	15.9	3.8	73	140	5.0	5.7	2.4
7.4	1010	601	21	7.5	23.7	5.7	49	85	7.1	8.6	2.4
8.0	969	258	50	7.8	22.6	7.4	46	64	9.7	10.3	3.1
7.1	980	303	38	7.0	20.8	6.2	51	81	8.1	8.9	3.1
7.8	1307	352	48	10.2	22.3	8.1	67	119	9.3	10.3	3.1
7.9	963	262	48	7.6	16.3	5.6	65	140	5.0	7.4	3.1
10.4	957	257	65	10.0	16.4	6.6	65	194	8.7	10.1	3.1
7.1	1109	290	45	7.9	19.4	6.4	63	160	5.5	7.8	3.1
7.3	1078	389	37	7.8	19.4	5.4	64	121	7.2	8.2	2.8
1.0	144	133	12	1.4	3.2	1.6	12	41	2.5	2.3	0.5
13.6%	13.4%	34.1%	32.4%	18.2%	16.5%	29.1%	18.3%	33.8%	35.5%	28.2%	16.9%

Appendix 9: Table 1 (continued)

Lung volume and mixing					Conventional indices													
APV [ml]	WO.eff. [%]	M0	M1	M2	V'_{CO_2} [ml/min]	V'_{O_2} [ml/min]	RQ	$F_{et\text{-}CO_2}$ [vol%]	$V_{d\text{-}Bohr}$ [ml]	$V_{D\text{-}phys}$ [ml]	V_D/V_T [%]	P_{aCO_2} [mmHg]	P_{aO_2} [mmHg]	P_{cO_2} [mmHg]	pH	Hb [g/100m]	$S_{\bar{v}O_2}$ [%]
1589	51.3	0.64	2.06	7.20	204	226	0.90	4.0	502	533	38	28.5	158.3	242	7.52	12.0	56.9
2611	43.0	0.66	1.94	7.07	219	281	0.78	5.0	401	498	41	39.0	141.0	220	7.44	13.0	63.0
2076	45.3	0.65	2.04	7.52	185	248	0.75	4.9	408	495	43	37.5	75.8	228	7.44	12.8	66.5
1388	73.6	0.62	1.69	4.56	181	236	0.77	4.6	216	315	29	36.0	161.3	226	7.40	12.2	50.4
1486	74.1	0.59	1.86	5.38	289	342	0.85	5.7	279	248	20	37.5	113.2	234	7.38	11.0	63.0
1345	67.6	0.73	1.57	4.70	225	282	0.80	5.2	273	283	27	36.0	103.5	237	7.40	11.1	50.9
2409	79.8	0.57	1.39	3.36	186	251	0.74	5.2	187	231	24	38.3	91.5	231	7.43	12.9	61.2
1577	79.0	0.72	1.38	3.60	168	208	0.81	4.3	284	305	25	30.8	74.3	248	7.47	11.7	64.9
1425	76.9	0.62	1.46	3.58	175	190	0.92	5.6	235	252	27	39.8	139.5	241	7.37	12.5	73.4
2223	73.6	0.59	1.46	3.77	164	199	0.82	5.7	191	219	25	41.3	107.3	230	7.37	13.2	78.6
2491	59.6	0.66	1.45	4.43	258	306	0.84	5.9	257	256	26	39.8	157.5	246	7.44	13.3	61.1
1410	42.3	0.59	1.77	5.90	176	201	0.88	5.5	203	219	26	38.3	84.0	235	7.46	11.7	66.4
1478	39.2	0.64	2.09	7.69	152	187	0.81	5.7	400	375	33	36.8	144.8	224	7.46	11.3	51.5
1925	42.4	0.63	2.00	7.24	173	216	0.80	4.5	380	397	33	30.8	108.8	234	7.55	12.8	63.3
1516	46.3	0.85	1.77	6.61	166	191	0.87	4.9	383	395	33	33.8	100.5	229	7.49	14.6	70.3
1704	56.3	0.73	1.74	6.02	177	204	0.87	4.8	360	404	34	33.8	184.5	231	7.47	11.8	54.3
1368	38.8	0.62	1.87	6.80	133	165	0.81	4.9	315	350	41	36.8	94.5	273	7.41	13.2	62.7
1257	66.1	0.70	1.82	5.78	211	254	0.83	5.7	452	516	43	42.0	153.0	238	7.34	10.5	52.0
2650	84.6	0.60	1.30	2.86	213	243	0.88	5.4	276	328	32	39.8	168.0	234	7.42	14.3	71.0
682	75.1	0.52	1.84	4.91	177	190	0.93	5.4	345	363	36	38.3	142.5	238	7.34	11.6	57.6
931	42.4	1.07	2.13	7.79	161	176	0.91	4.5	308	327	34	31.5	103.5	242	7.40	13.3	68.7
1430	71.6	0.60	1.76	5.30	164	171	0.96	4.5	223	239	24	31.5	174.0	240	7.47	13.1	67.0
2123	41.9	0.76	1.64	6.10	211	201	1.05	4.1	362	520	40	33.8	68.2	239	7.45	12.8	61.9
1983	34.7	1.04	2.09	8.53	214	231	0.93	5.5	256	288	30	39.0	135.8	228	7.38	13.5	57.4
3732	50.7	0.60	1.72	5.93	244	291	0.84	4.3	203	303	32	33.8	151.5	232	7.47	13.0	70.3
2648	64.3	0.83	1.41	3.67	215	227	0.95	5.2	291	338	30	37.5	128.3	227	7.38	11.0	67.4
1825	58.5	0.69	1.74	5.63	194	228	0.86	5.0	307	346	32	36.2	125.6	236	7.43	12.5	62.8
656	15.9	0.13	0.25	1.59	35	45	0.07	0.6	86	99	6	3.6	33.6	10	0.05	1.0	7.4
35.9%	27.1%	19.5%	14.5%	28.3%	18.1%	19.6%	8.5%	11.1%	27.9%	28.6%	19.9%	9.8%	26.8%	4.3%	0.7 %	8.3%	11.7%

Appendix 9: Table 1 (continued)

Exchange of CO_2 and O_2 — expected values

V_{ds} [ml]	M2-phase-2 [ml]	$Slope_{CO_2}$ [%/l]	alv.eff.$_{CO_2}$ [%]	Q'_S/Q'_T [%]	a-vDO_2 [ml/l]	CO [l/ml]	CI [l/min/m^2]	V_D/V_T [%]	VD-anat. [ml]	FRC [ml]
110	75	0.95	67.1	4.1	7.6	3.0	1.4	51	148	2044
203	88	1.42	71.1	5.1	6.9	4.1	2.2	48	148	2006
217	104	2.02	69.9	21.7	5.1	4.9	2.6	53	140	2031
138	42	0.51	80.5	3.0	8.8	2.7	1.3	53	158	2146
128	46	0.87	88.9	10.3	5.6	6.1	2.7	46	160	2047
114	48	1.05	81.7	8.0	7.6	3.7	1.8	56	175	2305
97	39	0.57	84.4	9.9	6.9	3.6	1.5	53	203	2424
144	56	0.62	84.7	15.4	5.4	3.9	2.1	53	130	1961
156	54	1.14	87.4	8.1	4.9	3.9	1.9	57	142	2109
139	47	0.90	88.9	13.7	4.0	5.0	2.7	51	140	1988
180	57	1.21	90.6	4.7	7.5	4.1	2.0	53	148	2087
119	41	1.26	86.4	13.5	5.3	3.8	2.3	48	109	1458
187	57	1.57	79.9	3.9	7.9	2.4	1.1	53	175	2268
164	53	0.91	77.6	7.8	6.6	3.3	1.6	55	146	2094
208	63	1.07	80.9	10.0	6.0	3.2	1.6	49	152	2045
196	61	1.03	78.5	2.2	8.0	2.6	1.3	50	150	2042
160	68	2.37	73.3	10.3	6.9	2.4	1.3	55	128	1617
337	77	1.07	79.6	5.0	7.2	3.5	1.9	55	146	2105
224	63	0.61	87.2	4.3	6.1	4.0	2.2	53	148	2081
193	58	1.72	79.0	5.4	7.1	2.7	1.4	56	130	2010
181	63	1.05	81.6	10.8	5.8	3.0	1.5	54	130	1628
121	39	0.78	85.9	3.8	6.4	2.7	1.6	52	132	1637
131	65	1.20	66.0	16.2	6.3	3.2	1.7	50	128	1894
161	57	1.19	84.1	4.9	8.2	2.8	1.4	55	152	2136
132	48	1.02	79.5	5.6	5.7	5.1	2.6	52	158	2140
192	55	0.97	83.8	7.7	5.2	4.4	2.4	51	146	2041
167	59	1.12	80.7	8.3	6.5	3.6	1.8	52	147	2013
50	15	0.43	6.7	4.8	1.2	0.9	0.5	3	19	217
30.0%	25.7%	38.4%	8.3%	57.6%	18.1%	25.9%	26.0%	5.3%	12.6%	10.8%

Appendix 9: Table 2.1 Volume Study

Patient	Disk Source	V_T ml	delta	V'_I ml/sec	delta	RR /min	delta	Temp. °C	delta	V'_{CO_2} ml/min	delta	F_{CO_2et} vol %	delta	PIE_{CO_2} ml	delta	$slope_{CO_2}$ vol %/l	delta
N	3b	1050		455		8.3		36.6		195		4.4		125		1.12	
	3c	731	319↓	450	5	11.9	−3.6	37.0	−0.4	183	12	4.7	−0.3	125	0	1.48	−0.36
	2a	1260	210	589	134	6.8	−1.5	36.2	−0.4	213	18	4.8	0.4	129	4	0.94	−0.18
A	1a, 2a	917		310		6.4		34.0		175		5.6		156		1.15	
	3	521	396↓	303	7	14.8	−8.4	34.6	−0.6	181	−6	5.4	0.2	125	31	2.40	−1.25
	4a, 5b	1359	442	307	−3	4.3	−2.1	34.8	0.8	174	−1	5.6	0.0	187	31	0.85	−0.30
Q	6a	869		256		9.0		36.0		165		4.9		131		1.13	
	6b	716	153↓	253	3	9.8	−0.8	36.9	−0.9	148	17	5.0	−0.1	122	9	1.52	−0.39
	5b	1057	188	264	8	7.2	−1.8	36.3	0.3	166	1	5.0	0.1	145	14	0.87	−0.26
C	4	1090		289		8.4		37.5		214		4.4		109		0.48	
	2	534	556↓	344	−55	12.5	−4.1	37.2	0.3	145	69	4.7	−0.3	98	11	1.32	−0.84
	3	1428	338	361	72	4.1	−4.3	37.3	−0.2	149	−65	4.6	0.2	128	19	0.36	−0.12
R	4b	1042		280		9.4		34.0		273		4.9		111		0.71	
	4c	731	311↓	274	6	14.4	−5.0	34.0	0.0	279	−6	4.9	0.0	118	−7	1.08	−0.37
E	1	842		392		10.1		34.0		236		5.6		108		1.33	
	2	455	387↓	388	4	21.8	−11.7	35.1	−1.1	200	36	5.5	0.1	113	−5	2.51	−1.18
	3	1396	554	392	0	5.9	−4.2	35.3	1.3	232	−4	5.7	0.1	145	37	0.85	−0.48
F	2	858		285		6.2		33.7		164		5.7		139		0.90	
	1	529	329↓	277	8	12.3	−6.1	34.4	−0.7	156	8	5.7	0.0	120	19	1.64	−0.74
	3	1211	353	287	2	4.2	−2.0	34.5	0.8	169	5	5.7	0.0	160	21	0.59	−0.31
S	1	857		275		8.1		37.6		161		5.2		162		0.91	
	2	472	385↓	302	−27	18.4	−10.3	38.1	−0.5	143	18	5.7	−0.5	155	7	2.53	−1.62
	3a, 4	1377	520	262	−13	4.9	−3.2	38.5	0.9	181	20	5.6	0.4	200	38	0.64	−0.27
mean ref		941		318		8.2		35.4		198		5.1		130		0.97	
sd ref		103		69		1.4		1.7		40		0.5		21		0.27	
mean delta			363		10		−4.6		0.0		8		0.0		15		−0.58
sd delta			123		43		3.3		0.7		28		0.2		14		0.45
mean small volume		586	355	324	−6	14.5	−6.3	35.9	−0.5	179	19	5.2	−0.1	122	8	1.81	−0.84
sd small volume		119	113	67	23	3.9	3.7	1.6	0.5	45	25	0.4	0.2	16	13	0.58	0.47
mean large volume		1298	372	352	29	5.3	−2.7	36.1	0.5	183	−4	5.3	0.2	156	23	0.7	−0.3
sd large volume		131	142	115	54	1.3	1.2	1.4	0.6	29	29	0.5	0.2	28	13	0.2	0.1

Appendix 9: Table 2.1 (continued)

alv.eff.-CO_2 %	delta	$F_{CO_2a\text{-}E}$ %	delta	PEEP cmH_2O	delta	$P_{awo\text{-}mean}$ cmH_2O	delta	C_L ml/cmH_2O	delta	C_{tot} ml/cmH_2O	delta	APV ml	delta	WO.eff. %	delta	CO l/min	delta
				0.0		5.5		81		60							
				0.0	0.0	4.7	0.8	70	11	52	8						
				0.0	0.0	4.9	−0.6	91	10	66	6						
87.4		0.1		0.0		4.0		80		52		1470		77.4		3.87	
87.2	0.2	0.4	−0.2	0.0	0.0	3.7	0.3	74	6	45	7	1392	78	76.1	1.4	3.56	0.3
81.6	− 5.8	0.2	0.1	0.0	0.0	4.8	0.8	92	12	60	8	1474	4	78.3	0.9	3.41	−0.5
				0.0		8.4		42		36							
		0.0		0.0	0.0	7.2	1.2	38	4	34	2						
		0.5		0.0	0.0	9.5	1.1	46	4	38	2						
				0.0								1881		75.8			
				0.0	0.0							1923	− 42	83.5	−7.7		
				0.0	0.0							1847	− 34	79.5	3.7		
82.4		0.4		0.0		7.8		82		54							
94.8	−12.4	0.0	0.4	0.0	0.0	6.2	1.6	79	3	49	5						
85.5		0.0		0.0		4.7		76		54		1565					
88.4	− 2.9	0.3	−0.3	0.0	0.0	4.1	0.6	63	13	44	10						
79.9	− 5.6	0.0	0.0	0.0	0.0	6.6	1.9	85	9	64	10	1576					
89.0		0.2		0.0		3.0		124		70		2023		73.6		5.08	
97.3	− 8.3	0.0	0.2	0.0	0.0	3.2	−0.2	108	16	60	10	1899	124	71.8	1.9	4.83	0.2
84.4	− 4.6	0.5	0.3	0.0	0.0	3.5	0.5	145	21	78	8	2040	17	79.8	6.2	5.20	0.2
89.2		0.1		0.0		5.4		75		55		1664				2.64	
88.4	0.8	0.5	−0.4	0.0	0.0	3.9	1.5	66	9	47	8	1531	133			2.62	0.0
87.6	− 1.6	0.0	−0.1	0.0	0.0	7.2	1.8	87	12	67	12	1735	71			3.19	0.7
86.7		0.2		0.0		5.5		80		54		1721		75.6		3.86	
2.8		0.2		0.0		2.0		24		10		228		1.9		1.22	
	− 4.5		0.0		0.0		0.9		10		7		44		1.0		0.2
	4.2		0.3		0.0		0.8		5		3		68		4.7		0.4
91.2	− 4.5	0.2	−0.1	0.0	0.0	4.7	0.8	71	9	47	7	1686	73	77.1	−1.5	3.67	0.2
4.5	5.7	0.2	0.4	0.0	0.0	1.5	0.7	21	5	8	3	266	81	6.0	5.4	1.11	0.1
83.4	− 4.4	0.2	0.1	0.0	0.0	6.1	0.9	91	11	62	8	1734	15	79.2	3.6	3.93	0.2
3.4	1,9	0.2	0.1	0.0	0.0	2.1	0.9	32	6	13	3	223	43	0.8	2.7	1.10	0.6

Appendix 9: Table 2.1 (continued)

Q'_S/Q'_T %	delta	avD_{O_2} vol%	delta	$P_{awo\text{-}max}$ cmH_2O	delta	$P_{awo\text{-}EI}$ cmH_2O	delta	R_{aw} $cmH_2O/l{\cdot}s$	delta	P_{aO_2} mmHg	delta	$V_{d\text{-}phys}$ ml	V_D/V_T %	delta	$V_{d\text{-}Bohr}$ ml	V_D/V_T %	P_{aCO_2} mmHg
				20.6		18.5		6.3									
				16.7	3.9	14.6	3.9	5.4	0.9								
				22	1.4	21.5	3.0	7	0.7								
8.1		4.9		19.1		18.5		3.9		139.5		251	27.4		235	25.6	40.18
11.5	−3.4	5.3	−0.4	12.5	6.6	11.8	6.7	3.8	0.1	105.0	34.5	177	34.0	− 6.6	154	29.6	40.25
7.3	−0.8	5.6	0.7	24.5	5.4	23.9	5.4	5	1.1	141.0	1.5	405	29.8	2.4	372	27.4	40.53
				24.2		23.8		10.8									
				21.6	2.6	21.1	2.7	11	−0.2								35.00
				28.1	3.9	27.9	4.1	11.2	0.4								38.29
				21.4		21.4		5.2		150.8		273	26.2		211	20.2	37.10
				16.1	5.3	16.1	5.3	4.7	0.5	134.3	16.5	148	20.2	6.0	162	22.2	34.30
				17.6		16.9		5.9		112.5		215	25.5		224	26.6	39.20
				12.3	5.3	11.6	5.3	5.6	0.3	101.3	11.2	152	33.4	− 7.9	137	30.1	40.39
				24.6	7.0	24.2	7.3	6	0.1	130.5	18.0	393	28.2	2.6	400	28.7	39.90
13.7		4.0		13.3		12.8		3.7		107.3		219	25.5		191	22.3	41.58
13.0	0.7	4.0	0.0	9.9	3.4	9.3	3.5	4.2	−0.5	111.0	− 3.7	140	26.5	− 0.9	130	24.6	39.90
18.5	4.8	3.8	−0.2	16.7	3.4	16.3	3.5	4.3	0.6	89.3	−18.0	322	26.6	1.1	246	20.3	43.33
5.8		7.9		17		16.9		5.1		105.9		236	27.5		227	26.5	36.89
5.7	0.1	7.1	0.8	11.3	5.7	11.3	5.6	4.2	0.9	130.5	−24.6	191	40.5	−12.9	168	35.6	43.26
3.8	−2.0	6.1	−1.8	22.4	5.4	22.4	5.5	5.5	0.4	177.0	71.1	343	24.9	− 2.6	392	28.5	39.20
9.2		5.6		19.0		18.4		5.8		123.2		239	26.4		218	24.2	39.33
4.1		2.0		3.5		3.5		2.4		20.6		24	1.0		17	2.8	2.55
	−0.1		−0.2		4.6		4.8		0.4		11.8			− 2.1			
	2.8		0.9		1.6		1.4		0.5		28.8			6.0			
10.1	−0.9	5.5	0.1	14.3	4.7	13.7	4.7	5.6	0.3	116.4	6.8	162	30.9	− 4.5	150	28.4	
3.9	2.2	1.6	0.6	4.0	1.4	4.0	1.4	2.5	0.5	15.1	22.2	21	7.8	7.2	16	5.2	
9.9	0.7	5.2	−0.4	23.1	4.4	22.7	4.8	6.5	0.6	134.5	18.2	366	27.4	0.9	353	26.6	
7.7	3.6	1.2	1.3	3.8	1.9	3.8	1.6	2.5	0.3	36.1	38.2	40	2.1	2.4	72	4.0	

Appendix 9: Table 2.2 PEEP Study

Patient	Disk Source	V_T ml	delta	V'_I ml/sec	delta	RR /min	delta	Temp. °C	delta	V'_{CO_2} ml/min	delta	$F_{CO_2\text{-et}}$ vol %	delta	PIE_{CO_2} ml	delta	$slope_{CO_2}$ vol %/l	delta
A	8	925		312		5.8		35.2		165		5.8		159		1.32	
	9	940	15	321	9	7.8	2.0	35.3	0.1	196	31	5.4	−0.4	182	23	1.03	−0.29
	10	945	5	325	4	6.7	−1.1	35.5	0.2	174	−22	5.8	0.4	188	6	1.01	−0.02
B	1	1218		227		5.8		34.6		170		4.4		143		0.63	
	2	1160	−58	218	− 9	5.7	−0.1	34.8	0.2	178	8	4.8	0.4	138	− 5	0.46	−0.17
C	4	1090		289		8.4		37.5		214		4.4		109		0.48	
	5	1089	− 1	355	66	9.3	0.9	37.6	0.1	205	− 9	4.2	−0.2	110	1	0.28	−0.20
	6	1072	−17	364	9	8.3	−1.0	37.6	0.0	175	−30	4.3	0.1	122	12	0.40	0.12
D	2	1033		493		7.8		35.8		225		5.2		114		1.05	
	3d, 3e	1026	− 7	482	−11	7.9	0.1	36.2	0.4	222	− 3	5.2	0.0	136	22	0.64	−0.41
E	7	855		480		10.6		36.8		235		5.6		128		1.19	
	8	830	−25	468	−12	10.5	−0.1	36.9	0.1	230	− 5	5.7	0.1	155	27	1.10	−0.09
	9	891	61	436	−32	10.0	−0.5	37.2	0.3	259	29	5.7	0.0	171	16	0.97	−0.13
F	7, 8	853		309		6.7		35.5		182		5.8		137		0.95	
	9, 10	850	− 3	306	− 3	6.7	0.0	35.7	0.2	176	− 6	5.9	0.1	163	26	0.93	−0.02
	11	838	−12	307	1	8.1	1.4	36.1	0.4	194	18	5.9	0.0	195	32	1.05	0.12
G	1	1038		430		7.6		36.2		213		5.4		224		0.61	
	2	1055	17	415	−15	7.5	−0.1	36.6	0.4	221	8	5.8	0.4	264	40	0.64	0.03
mean ref		1002		363		7.5		35.9		201		5.2		145		0.89	
sd ref		133		104		1.7		1.0		28		0.6		39		0.32	
mean delta			− 2		1		0.1		0.2		2		0.1		18		−0.10
sd delta			29		25		1.0		0.1		19		0.3		14		0.17
mean ZEEP-PEEP		993	− 9	366	4	7.9	0.4	36.2	0.2	204	3	5.3	0.1	164	19	0.73	−0.16
sd ZEEP-PEEP		124	26	95	29	1.6	0.8	1.0	0.1	22	14	0.6	0.3	50	16	0.31	0.15
mean PEEP-PEEP		937	9	358	− 5	8.3	−0.3	36.6	0.2	201	− 1	5.4	0.1	169	17	0.86	0.02
sd PEEP-PEEP		100	36	57	19	1.4	1.2	1.0	0.2	40	29	0.8	0.2	33	11	0.31	0.12

Appendix 9: Table 2.2 (continued)

alv.eff.-CO_2 %	delta	$F_{CO_2a\text{-}E}$ vol %	delta	PEEP cmH_2O	delta	$P_{awo\text{-}mean}$ cmH_2O	delta	C_L ml/cmH_2O	delta	C_{tot} ml/cmH_2O	delta	APV ml	delta	WO.eff. %	delta	CO l/min	delta
83.9		0.5		0.0		3.8		76		51		1385		73.1		4.62	
82.4	−1.5	0.5	0.0	6.3	6.3	10.6	6.8	83	7	56	5	1770	385	68.9	− 4.2	3.95	−0.4
85.9	3.5	0.2	−0.3	11.1	4.8	15.4	4.8	90	7	55	−1	2551	781	55.6	−13.3	3.60	−0.5
84.8		0.1		0.0		5.9		106		73		1577		79.0		4.18	
85.9	1.1	0.2	0.1	6.1	6.1	12.5	6.6	132	26	71	−2	2222	645	74.5	− 4.6	3.25	−0.3
				0.0								1881		75.8			
				6.0	6.0							2645	764	76.6	0.7		
				12.0	6.0							3412	767	73.0	− 3.6		
81.6		0.1		0.0		4.6		82		60		1345		67.6		3.71	
82.0	0.4	0.2	0.1	7.1	7.1	11.4	6.8	83	1	65	5	2172	827	73.7	6.1	3.84	0.1
81.2		0.0		0.0		4.4		76		57		1452		63.0			
86.5	5.3	0.3	0.3	4.8	4.8	9.1	4.7	76	0	59	2	1810	358	49.7	−13.3		
88.4	1.9	0.3	0.1	10.7	5.9	15.3	6.2	78	2	57	−2	2700	890	64.4	14.7		
89.1		0.3		0.0		3.1		154		72		1974		72.4		6.25	
87.9	−1.2	0.5	0.2	5.5	5.5	8.0	4.9	154	0	78	6	2517	543	78.8	6.5	6.50	0.2
85.2	−2.7	0.8	0.3	10.7	5.2	13.9	5.9	124	−30	74	−4	3207	690	65.7	−13.2	5.32	−0.8
87.2		0.4		0.0		3.8		140		73		2650		84.6		3.94	
87.5	0.3	0.5	0.1	6.8	6.8	10.5	6.7	130	−10	77	4	3379	729	76.4	− 8.2	3.78	−0.2
84.6		0.2		0.0		4.3		106		64		1752		73.6		4.54	
3.1		0.2		0.0		1.0		34		10		464		7.1		1.01	
	0.8		0.1		5.9		5.9		0		1		671		− 2.9		−0.3
	2.5		0.2		0.7		0.9		15		4		174		9.2		0.4
85.4	0.7	0.4	0.1	6.1	6.1	10.4	6.1	110	4	68	3	2359	607	71.2	− 2.4	4.27	−0.1
2.6	2.5	0.2	0.1	0.8	0.8	1.6	1.0	33	12	9	3	555	185	10.0	7.3	1.28	0.3
86.5	0.9	0.4	0.0	11.1	5.5	14.9	5.6	97	− 7	62	−2	2968	782	64.7	− 3.8	4.46	−0.7
1.7	3.2	0.3	0.3	0.6	0.6	0.8	0.7	24	20	10	2	408	82	7.1	13.2	1.21	0.2

Appendix 9: Table 2.2 (continued)

Q'_S/Q'_T %	delta	avD_{O_2} vol%	delta	$P_{awo\text{-}max}$ cmH_2O	delta	$P_{awo\text{-}EI}$ cmH_2O	delta	R_{aw} $cmH_2O/l\cdot s$	delta	P_{aO_2} mmHg	delta	$V_{d\text{-}phys}$ ml	V_D/V_T %	delta	$V_{d\text{-}Bohr}$ ml	V_d/V_t %	P_{aCO_2} mmHg
7.5		4.5		19.4		18.7		5.0		142.5		282.0	30.5		230.0	24.9	44.0
5.3	−2.2	5.2	0.7	25.2	5.8	24.6	5.9	4.6	−0.4	160.5	18.0	317.0	33.7	3.2	259.0	27.6	41.3
4.6	−0.7	5.8	0.6	30.9	5.7	30.3	5.7	4.4	−0.2	162.0	1.5	296.0	31.3	−2.4	271.0	28.7	42.1
15.4		6.4		17.7		17.3		6.3		74.3		305.0	25.0		287.0	23.6	31.4
12.9	−2.5	5.4	−1.0	24.2	6.5	23.8	6.5	6.2	−0.1	80.3	6.0	278.0	24.0	−1.1	241.0	20.8	35.1
8.0		7.6		18.6		17.7		7.7		103.5		284.0	27.5		272.0	26.3	37.1
6.5	−1.5	7.2	−0.4	24.9	6.3	24.4	6.7	5.9	−1.8	124.5	21.0	296.0	28.8	1.4	262.0	25.5	38.1
				16.8		15.3		7.0		135.0		260.0	30.4		221.0	25.8	39.2
				20.8	4.0	19.9	4.6	5.3	−1.7	141.0	6.0	246.0	29.6	−0.8	220.0	26.5	41.7
				28.4	7.6	27.9	8.0	4.6	−0.7	137.3	− 3.7	257.0	28.8	−0.8	223.0	25.0	42.0
16.9		3.6		12.9		12.2		4.1		95.3		215.0	25.2		185.0	21.7	42.6
16.1	−0.8	3.4	−0.2	17.9	5.0	17.3	5.1	4.2	0.1	101.3	6.0	246 .0	28.9	3.7	194.0	22.8	44.9
12.7	−3.4	4.0	0.6	24.2	6.3	23.8	6.5	4.4	0.2	111.8	10.5	289.0	34.5	5.5	217.0	25.9	46.8
4.3		6.1		15.9		15.5		5.0		168.0		328.0	31.6		276.0	26.6	40.6
2.7	−1.6	6.9	0.8	22.6	6.7	22.0	6.5	4.3	−0.7	184.5	16.5	364.0	34.5	2.9	302.0	28.6	44.2
10.4		5.6		16.9		16.1		5.9		119.8		279.0	28.4		245.2	24.8	40.73
5.4		1.6		2.3		2.3		1.4		34.7		38.9	2.9		39.6	1.9	40.13
	−1.8		0.2		6.0		6.2		−0.6		9.1			1.3			
	1.0		0.7		1.0		1.0		0.7		8.1			2.7			
8.7	−1.7	5.6	0.0	22.6	5.7	22.0	5.9	5.1	−0.8	132.0	12.3	291.2	29.9	1.6	246.3	25.3	
5.6	0.7	1.5	0.8	2.8	1.0	2.9	0.9	0.8	0.8	38.3	7.0	45.3	3.8	2.1	37.3	3.0	
8.7	−2.1	4.9	0.6	27.8	6.5	27.3	6.7	4.5	−0.2	137.0	2.8	280.7	31.6	0.8	237.0	26.5	
5.7	1.9	1.3	0.0	3.4	1.0	3.3	1.2	0.1	0.5	25.1	7.2	20.8	2.8	4.2	29.6	1.9	

Appendix 9: Table 2.3 Inspiration Flow Study

Patient	Disk Source	V_T ml	delta	V'_I ml/sec	delta	RR /min	delta	Temp. °C	delta	V'_{CO_2} ml/min	delta	$F_{CO_2\text{-et}}$ vol %	delta	PIE_{CO_2} ml	delta	$slope_{CO_2}$ vol %/l	delta
N	2b	1241		609		6.9		36.0		190		4.3		152		0.93	
	3a	1236	– 5	960	351	7.1	0.2	36.1	0.1	180	−10	4.2	−0.1	168	16	0.96	0.03
H	1	1205		365		6.9		35.3		173		4.5		165		0.91	
	2	1236	31	660	295	6.9	0.0	35.4	0.1	170	– 3	4.6	0.1	192	27	0.98	0.07
I	2, 3	1195		406		6.1		35.2		169		4.9		184		1.15	
	4.0	1176	– 19	790	384	6.0	−0.1	35.6	0.4	145	−24	4.9	0.0	187	3	1.23	0.08
J	8	1180		357		6.9		36.7		177		4.8		196		1.03	
	7b	1200	20	801	444	6.1	−0.8	36.5	−0.2	147	−30	4.9	0.1	236	40	1.24	0.21
D	3	1060		262		7.5		39.3		186		4.9		172		1.64	
	6	1160	100	617	355	7.6	0.1	39.3	0.0	185	– 1	4.9	0.0	217	45	2.02	0.38
P	1	856		394		7.4		35.0		176		5.5		119		1.26	
	2	830	– 26	682	288	8.5	1.1	35.5	0.5	150	−26	5.3	−0.2	167	48	1.92	0.66
O	2	1130		337		5.9		35.5		153		5.7		187		1.57	
	3	1140	10	712	375	5.9	0.0	35.8	0.3	149	– 4	5.7	0.0	236	49	1.74	0.17
M	1	1185		361		6.2		35.0		166		4.9		208		1.07	
	2	1200	15	706	345	6.4	0.2	35.5	0.5	157	– 9	5.1	0.2	213	5	1.29	0.22
mean ref		1132		386		6.7		36.0		174		4.9		173		1.20	
sd ref		124		100		0.6		1.5		12		0.5		28		0.28	
mean delta			16		355		0.1		0.2		−13		0.0		29		0.23
sd delta			39		50		0.5		0.3		12		0.1		19		0.21
mean high flow		1147		741		6.8		36.2		160		5.0		202		1.42	
sd high flow		133		108		0.9		1.3		16		0.4		28		0.41	

Appendix 9: Table 2.3 (continued)

alv.eff.-CO_2 %	delta	$F_{CO_{2a-E}}$ vol %	delta	PEEP cmH_2O	delta	$P_{awo-mean}$ cmH_2O	delta	C_L ml/cmH_2O	delta	C_{tot} ml/cmH_2O	delta	APV ml	delta	WO.eff. %	delta	CO l/min	delta
				0.0		4.2		86		63							
				0.0	0.0	4.0	−0.2	79	− 7	59	−4						
77.7		0.1		0.0		5.6		123		67		1925		42.4		3.24	
77.2	− 0.5	0.1	0.0	0.0	0.0	4.3	−1.3	127	4	67	0	1688	−237	54.1	11.7	3.63	0.4
70.9		0.7		0.0		3.4		144		92		2106		62.1		4.49	
74.6	3.7	0.3	−0.4	0.0	0.0	2.8	−0.6	127	−17	86	−6	1669	−437	56.4	− 5.6	3.51	−1.0
78.5		0.3		0.0		4.5		133		86						2.56	
71.2	− 7.3	0.7	0.4	0.0	0.0	2.8	−1.7	123	−10	81	−5					2.09	−0.5
85.8		0.0		0.0		5.1		150		82		2400		47.1		5.18	
69.8	−16.0	0.7	0.7	0.0	0.0	4.6	−0.5	155	5	85	3	2076	−324	45.3	− 1.9	4.83	−0.3
86.4		0.1		0.0		5.2		69		40		1410		42.3		3.77	
81.7	− 4.7	0.2	0.1	0.0	0.0	4.4	−0.8	63	− 6	37	−3	1183	−227	37.6	− 4.7	2.93	−0.8
79.7		0.0		0.0		5.2		91		62		1478		39.2		2.36	
74.6	− 5.1	0.2	0.2	0.0	0.0	2.6	−2.6	72	−19	63	1	1277	−201	38.7	− 0.5	2.16	−0.2
80.9		0.1		0.0		5.4		98		63		1516		46.3		3.17	
75.8	− 5.1	0.2	0.1	0.0	0.0	4.2	−1.2	91	− 7	63	0	1534	18	53.8	7.5	3.26	0.1
80.0		0.2		0.0		4.8		112		69		1806		46.6		3.54	
5.3		0.3		0.0		0.7		30		17		401		8.1		1.02	
	− 5.0		0.1		0.0		−1.1		− 7		−2		−235		1.1		−0.3
	6.1		0.3		0.0		0.8		9		3		151		7.0		0.5
75.0		0.3		0.0		3.7		105		68		1571		47.6		3.20	
3.9		0.2		0.0		0.8		33		16		321		8.3		0.94	

Appendix 9: Table 2.3 (continued)

Q'_S/Q'_T %	delta	avD_{O_2} vol%	delta	$P_{awo\text{-}max}$ cmH_2O	delta	$P_{awo\text{-}El}$ cmH_2O	delta	R_{aw} $cmH_2O/l \cdot s$	delta	P_{aO_2} mmHg	delta	$V_{d\text{-}phys}$ ml	V_D/V_T %	delta	$V_{d\text{-}Bohr}$ ml	V_d/V_t %	P_{aCO_2} mmHg
				23.1		21.5		6.8									
				26.6	3.5	22.8	1.3	6.9	0.1								
7.8		6.6		21.6		20.3		8.5		108.8		397	32.9		381	31.6	32.13
7.3	−0.5	5.7	−0.9	23.9	2.3	22.0	1.7	7.5	−1.0	127.5	18.7	429	34.7	1.8	418	33.8	32.62
4.2		4.4		15.6		14.3		5.7		176.3		480	40.2		371	31.0	39.41
4.7	0.5	5.2	0.8	19.3	3.7	16.2	1.9	8.0	2.3	163.5	−12.8	434	36.9	− 3.3	388	33.0	36.47
2.2		8.0		16.2		15.2		5.8		184.5		404	34.2		361	30.6	35.49
2.4	0.2	8.6	0.6	20.3	4.1	17.6	2.4	6.4	0.6	184.5	0.0	510	42.5	8.3	411	34.3	39.27
24.0		4.7		13.9		13.4		3.8		71.3		298	28.1		305	28.8	34.3
21.7	−2.3	5.1	0.4	16.9	3.0	15.0	1.6	4.0	0.2	75.8	4.5	494	42.6	14.5	408	35.2	38.85
13.5		5.3		25.0		23.4		12.4		84.0		219	25.6		204	23.8	39.41
10.6	−2.9	5.8	0.5	30.0	5.0	25.0	1.6	13.5	1.1	98.3	14.3	286	34.5	8.9	268	32.3	38.43
3.9		7.9		20.7		19.5		10.7		144.8		375	33.2		400	35.4	39.9
2.9	−1.0	8.1	0.2	22.9	2.2	20.2	0.7	9.0	−1.7	178.5	33.7	472	41.4	8.2	445	39.0	41.44
10.0		6.0		21.7		20.5		10.4		100.5		395	33.3		383	32.3	34.79
8.4	−1.6	5.5	−0.5	25.2	3.5	21.3	0.8	9.7	−0.7	119.3	18.8	448	37.3	4.0	422	35.2	36.96
9.4		6.1		19.7		18.5		8.0		124.3		367	32.5		344	30.5	37.11
7.5		1.4		4.0		3.7		3.0		44.7		84	4.7		69	3.6	2.88
	−1.1		0.2		3.4		1.5		0.1		11.0			6.0			
	1.3		0.6		0.9		0.6		1.3		15.1			5.7			
8.3		6.3		23.1		20.0		8.1		135.3	11.0	439	38.6		394	34.7	
6.6		1.4		4.2		3.5		2.8		41.4	15.1	74	3.6		58	2.2	

Appendix 9: Table 2.4 EIP Study

Patient	Disk Source	V_T ml	delta	V'_I ml/sec	delta	RR /min	delta	Temp. °C	delta	V'_{CO_2} ml/min	delta	$F_{CO_2\text{-et}}$ vol %	delta	PIE_{CO_2} ml	delta	$slope_{CO_2}$ vol%/l	delta
H	2	1236		660		6.9		35.4		170		4.6		192		0.98	
	4	1211	−25	660	0	6.3	−0.6	36.2	0.8	173	3	4.7	0.1	150	−42	0.72	−0.26
I	4	1176		790		6.0		35.6		145		4.9		187		1.23	
	6	1216	40	790	0	5.5	−0.5	36.4	0.8	147	2	4.8	−0.1	174	−13	0.79	−0.44
K	2	1050		339		9.1		36.4		254		5.5		191		1.25	
	4	1060	10	339	0	9.9	0.8	37.1	0.7	284	30	5.2	−0.3	175	−16	0.75	−0.50
L	3	1140		712		5.9		35.8		149		5.7		236		1.74	
	6	1155	15	712	0	6.8	0.9	36.3	0.5	168	19	5.5	−0.2	185	−51	1.22	−0.52
M	3	1223		722		6.4		36.4		161		5.2		247		1.25	
	5	1245	22	722	0	6.3	−0.1	37.0	0.6	179	18	5.0	−0.2	192	−55	0.79	−0.46
F	7, 8	855		309		6.2		35.5		182		5.8		137		0.95	
	5	853	− 2	309	0	6.7	0.5	34.7	−0.8	183	1	5.8	0.0	84	−53	0.43	−0.52
mean ref		1113		589		6.8		35.9		177		5.3		198		1.23	
sd ref		143		209		1.2		0.4		40		0.5		39		0.28	
mean delta			10		0		0.2		0.4		12		−0.1		−38		−0.45
sd delta			22		0		0.7		0.6		12		0.1		19		0.10
mean hold		1123		589		6.9		36.3		189		5.2		160		0.78	
sd hold		148		209		1.5		0.9		48		0.4		40		0.25	

Appendix 9: Table 2.4 (continued)

alv.eff.-CO_2 %	delta	$F_{CO_{2a\text{-}E}}$ vol %	delta	PEEP cmH_2O	delta	$P_{awo\text{-}mean}$ cmH_2O	delta	C_L ml/cmH_2O	delta	C_{tot} ml/cmH_2O	delta	APV ml	delta	WO.eff. %	delta	CO l/min	delta
77.2		0.1		0.0		4.3		127		67		1688		54.1		3.63	
80.6	3.4	0.2	0.1	0.0	0.0	5.6	1.3	131	4	70	3	1780	92	58.4	4.3	3.96	0.3
74.6		0.3		0.0		2.8		127		86		1669		56.4		3.51	
75.3	0.7	0.6	0.3	0.0	0.0	4.2	1.4	136	9	83	−3	1463	−206	51.3	−5.1	1.89	−1.5
86.7		0.2		0.0		5.5		126		70		2335		56.5		4.03	
91.4	4.7	0.1	0.0	0.0	0.0	8.2	2.7	133	7	76	6	2768	433	62.2	5.7	4.65	0.6
74.6		0.2		0.0		2.6		72		63		1277		39.7		2.16	
76.9	2.3	0.0	−0.2	0.0	0.0	6.0	3.4	70	−2	61	−2	1121	−156	29.0	−10.7	2.67	0.6
72.5		0.7		0.0		4.1		106		66		1572		58.6		3.96	
76.8	4.3	0.5	−0.2	0.0	0.0	5.7	1.6	106	0	67	1	1529	−43	53.6	−5.0	3.39	−0.5
89.1		0.3		0.0		3.1		154		72		1974		72.4		6.25	
93.8	4.7	0.1	−0.2	0.0	0.0	4.8	1.7	155	1	79	7	2109	135	72.2	−0.2	5.64	−0.5
79.1		0.3		0.0		3.7		119		71		1753		56.3		3.92	
7.0		0.2		0.0		1.1		27		8		363		10.4		1.33	
	3.4		0.0		0.0		2.0		3		2		43		−1.8		−0.2
	1.6		0.2		0.0		0.8		4		4		233		6.3		0.8
82.5		0.2		0.0		5.8		122		73		1795		54.5		3.70	
8.1		0.2		0.0		1.4		30		8		580		14.5		1.36	

Appendix 9: Table 2.4 (continued)

Q'_S/Q'_T %	delta	avD_{O_2} vol%	delta	$P_{awo\text{-}max}$ cmH_2O	delta	$P_{awo\text{-}EI}$ cmH_2O	delta	R_{aw} $cmH_2O/l{\cdot}s$	delta	P_{aO_2} mmHg	delta	$V_{d\text{-}phys}$ ml	V_D/V_T %	delta	$V_{d\text{-}Bohr}$ ml	V_d/V_t %
7.3		5.7		23.9		22.0		7.5		127.5		429	34.7		418	33.8
6.1	−1.2	5.4	−0.3	20.9	−3.0	17.6	−4.4	6.5	−1.0	144.8	17.3	354	29.2	−5.5	319	26.3
4.7		5.2		19.2		16.2		8.0		163.5		434	36.9		388	33.0
2.3	−2.4	9.5	4.3	19.0	−0.2	15.0	−1.2	6.6	−1.4	171.0	7.5	424	34.9	−2.0	334	27.5
4.9		7.1		16.8		16.1		5.4		153.8		302	28.8		279	26.6
4.4	−0.5	6.8	−0.3	15.1	−1.7	14.5	−1.6	4.9	−0.5	172.5	18.7	249	23.5	−5.3	230	21.7
2.9		8.1		22.9		20.2		9.0		178.5		472	41.4		445	39.0
3.6	0.7	6.6	−1.5	23.0	0.1	18.9	−1.3	11.0	2.0	163.5	−15.0	408	35.3	−6.1	424	36.7
10.4		4.8		24.4		21.4		7.9		111.8		510	41.7		421	34.4
8.6	−1.8	6.0	1.2	22.3	−2.1	18.6	−2.8	8.0	0.1	114.8	3.0	430	34.5	−7.2	351	28.2
16.9		3.6		12.9		12.2		4.1		95.3		215	25.1		185	21.6
16.3	−0.6	3.8	0.2	11.3	−1.6	10.5	−1.7	4.0	−0.1	93.7	− 1.6	132	15.5	−9.7	122	14.3
7.9		5.8		20.0		18.0		7.0		138.4		394	34.8		356	31.4
5.1		1.6		4.6		3.8		1.8		32.1		112	6.7		102	6.2
	−1.0		0.6		−1.4		−2.2		−0.2		5.0			−5.9		
	1.1		2.0		1.2		1.2		1.2		12.6			2.5		
6.9		6.4		18.6		15.9		6.8		143.4		333	28.8		297	25.8
5.1		1.9		4.6		3.2		2.5		32.6		119	8.0		106	7.4

Appendix 9: Table 3.1. Description of patients

Patient	Age years	Weight kg	Height cm	Surface m^2	FRC expected liters
SRH	36.0	83.5	180.0	2.06	3.35
SIJ	68.0	83.0	186.0	2.08	4.17
RBR	61.0	69.0	164.0	1.79	2.82
SCG	59.0	111.0	198.0	2.48	4.70
UGE	67.0	74.4	172.0	1.90	3.35
RLS	50.0	81.5	175.0	2.01	3.29
ZGR	60.0	68.9	173.0	1.83	3.31
JCO	60.0	74.4	164.0	1.86	2.81
mean	57.6	80.7	176.5	2.0	3.5
SD	10.3	13.6	11.4	0.2	0.6

Appendix 9: Table 3.2. Gas exchange during the CPPV period

Patient	V_T ml	RR /min	V'_{CO_2} ml/min	$F_{CO_2\text{-et}}$ vol %	V_{dS} ml	$slope_{CO_2}$ vol %/l	alv.eff.-CO_2 %	$P_{awo\text{-}EE}$ cmH_2O	$P_{awo\text{-}max}$ cmH_2O
SRH	1278.0	10.0	312.0	4.8	80.0	1.30	73.0	6.4	33.9
SIJ	1022.0	13.1	284.0	3.8	160.0	0.69	75.2	6.4	23.8
RBR	1160.0	5.7	178.0	4.8	138.0	0.47	85.9	6.1	24.4
SCG	968.0	5.9	185.0	5.7	123.0	0.26	86.0	4.9	20.8
UGE	1092.5	9.2	229.0	5.0	274.0	0.70	69.6	7.7	22.0
RLS	1107.0	8.4	271.0	5.2	213.0	1.05	77.9	7.6	27.7
ZGR	1055.0	7.5	221.0	5.8	264.0	0.64	87.5	8.3	24.1
JCO	1090.5	7.0	199.0	5.3	172.0	0.85	82.3	5.3	28.1
mean	1096.6	8.4	234.9	5.1	178.0	0.75	79.7	6.6	25.6
SD	93.3	2.4	49.1	0.6	68.1	0.32	6.7	1.2	4.2

Patient	$P_{awo\text{-}mean}$ cmH_2O	APV ml	CO l/min	Q'_S/Q'_T %	a-vDO_2 vol%	P_{aCO_2} kPa	P_{aO_2} kPa	MV l/min	V_D/V_T
SRH	12.6	2499.0	5.9	4.4	5.6	4.5	23.2	12.8	0.39
SIJ	9.5	1949.0	3.7	6.4	7.9	4.1	16.6	13.4	0.42
RBR	12.5	2224.0	3.3	13.0	6.4	4.6	10.7	6.6	0.34
SCG	7.9	3277.0	4.0	11.7	5.9	5.4	11.6	5.7	0.32
UGE	12.1	1958.0	3.7	6.7	7.0	6.1	18.6	10.1	0.58
RLS	13.2	2238.0	4.2	9.7	7.3	5.6	13.3	9.3	0.41
ZGR	12.0	3351.0	3.6	2.7	6.9	5.8	24.5	7.9	0.46
JCO	13.2	1270.0	3.4	6.4	6.3	5.2	18.8	7.6	0.43
mean	11.6	2345.8	4.0	7.6	6.7	5.2	17.1	9.2	0.42
SD	1.9	696.2	0.8	3.6	0.8	0.7	5.1	2.8	0.08

Appendix 9: Table 3.3. Gas exchange during the IMV period

Patient	V'_{CO_2} ml/min	$P_{awo\text{-}EE}$ cmH_2O	$P_{awo\text{-}mean}$ cmH_2O	APV ml	CO ml/min	Q'_S/Q'_T %	a-vDO_2 vol%	P_{aCO_2} kPa	P_{aO_2} kPa	MV l/min	V_D/V_T
SRH	229.0	4.6	6.9	2327.0	5.8	2.6	4.8	5.7	23.6	8.3	0.45
SIJ	297.0	6.6	6.6	2416.0	4.8	6.6	6.6	5.7	17.6	8.7	0.33
RBR	220.0	4.2	7.7	1888.0	4.7	6.1	5.3	5.7	20.1	7.3	0.40
SCG	316.0	5.3	6.7	2469.0	6.0	15.7	5.3	6.2	11.4	8.6	0.33
UGE	165.0	9.8	9.0	1217.0	5.0	19.0	4.6	4.7	12.0	15.5	0.74
RLS	237.0	6.3	6.7	1487.0	4.4	9.1	6.0	6.6	16.1	8.8	0.54
ZGR	250.0	11.0	12.1	4241.0	5.6	2.2	5.0	6.2	26.2	13.5	0.66
JCO	260.0	9.0	11.2	1294.0	5.6	11.3	4.7	6.7	16.3	7.2	0.39
mean	246.8	7.1	8.4	2167.4	5.2	9.1	5.3	5.9	17.9	9.7	0.5
SD	46.8	2.5	2.2	977.3	0.6	6.0	0.7	0.6	5.2	3.0	0.2

Appendix 9: Table 3.4. Gas exchange during the CPAP period

Patient	V_T ml	RR /min	V'_{CO_2} ml/min	$F_{CO_2\text{-}et}$ vol %	V_{dS} ml	$slope_{CO_2}$ vol %/l	alv.eff.-CO_2 %	$P_{awo\text{-}EE}$ cmH_2O	$P_{awo\text{-}max}$ cmH_2O
SRH	427.0	22.0	259.0	6.1	93.0	3.03	84.8	5.0	5.5
SIJ	510.0	18.5	299.0	6.7	87.0	4.30	96.1	3.4	3.8
RBR	585.0	12.8	210.0	6.1	103.0	2.60	86.7	6.3	6.3
SCG	768.0	12.0	322.0	6.7	103.0	1.10	85.7	6.3	6.5
UGE	777.0	17.9	175.0	5.1	283.0	1.98	77.1	8.4	8.8
RLS	870.0	10.2	222.0	6.1	225.0	2.55	74.2	8.3	8.3
ZGR	591.0	19.4	225.0	6.0	211.0	3.53	92.5	10.5	10.8
mean	646.9	16.1	244.6	6.1	157.9	2.73	85.3	6.9	7.1
SD	160.9	4.1	51.8	0.5	79.8	1.04	7.8	2.4	2.3

Patient	$P_{awo\text{-}mean}$ cmH_2O	APV ml	CO l/min	Q'_S/Q'_T %	a-vDO_2 vol%	P_{aCO_2} kPa	P_{aO_2} kPa	MV l/min	V_D/V_T
SRH	5.1	2027.0	4.6	3.1	6.6	6.1	21.9	9.4	0.49
SIJ	3.4	2302.0	5.5	6.9	5.9	5.7	17.9	9.4	0.37
RBR	5.8	1958.0	4.4	7.1	5.3	5.8	18.0	7.5	0.45
SCG	5.8	2441.0	7.1	21.3	5.3	6.4	9.7	9.2	0.38
UGE	8.1	1024.0	4.9	11.3	5.1	5.6	16.0	13.9	0.75
RLS	7.6	1259.0	4.3	12.4	5.7	6.4	13.7	8.9	0.56
ZGR	10.2	3501.0	4.7	2.0	5.6	5.7	26.5	11.5	0.61
mean	6.6	2073.1	5.1	9.2	5.6	5.9	17.7	10.0	0.52
SD	2.2	817.0	1.0	6.6	0.5	0.3	5.5	2.1	0.13

Appendix 9: Table 3.5. Breath-by-breath indices of gas exchange during IMV

Patient	Breath	ΔV_T ml	V_T ml	V_T Δ	RR /min	RR Δ	$F_{CO_2\text{-et}}$ vol %	$F_{CO_2\text{-et}}$ Δ	V_{dS} ml	V_{dS} Δ
SRH	MB	0.0	1203.0		2.9		6.0		114.0	
	SB		460.0	−743.0	8.8	5.9	6.3	0.3	86.0	− 28.0
SIJ	MB	−127.0	1342.5		3.3		6.0		142.0	
	SB		597.0	−745.5	6.6	3.3	6.4	0.4	108.0	− 34.0
RBR	MB	23.0	942.5		4.1		6.1		114.0	
	SB		445.0	−497.5	7.8	3.7	6.1	0.0	79.0	− 35.0
SCG	MB	− 15.0	1060.0		4.8		6.3		81.0	
	SB		389.0	−671.0	9.0	4.2	6.7	0.4	37.0	− 44.0
UGE	MB	183.0	1090.0		4.2		4.8		409.0	
	SB		737.0	−353.0	14.3	10.1	4.8	0.0	303.0	−106.0
RLS	MB	100.0	1305.0		4.1		6.0		249.0	
	SB		696.5	−608.5	4.8	0.7	6.1	0.1	165.0	− 84.0
ZGR	MB	− 15.0	1339.5		3.8		6.0		381.0	
	SB		561.0	−778.5	15.1	11.3	6.2	0.2	239.0	−142.0
JCO	MB	19.0	1090.0		3.2		6.1		86.0	
	SB		464.5	−625.5	5.1	1.9	6.1	0.0	87.0	1.0
mean MB		21.0	1171.6		3.8		5.9		197.0	
SD MB		90.6	148.5		0.6		0.5		133.1	
mean Δ				−627.8		5.1		0.2		− 59.0

Patient	Breath	$slope_{CO_2}$ vol %	$slope_{CO_2}$ Δ	alv.eff.-CO_2 %	alv.eff.-CO_2 Δ	V'_A l/min	V'_A Δ	$P_{awo\text{-}max}$ cmH_2O	$P_{awo\text{-}max}$ Δ
SRH	MB	1.15		82.2		3.16		34.0	
	SB	2.98	1.83	94.4	12.2	3.29	0.13	6.0	−28.0
SIJ	MB	0.74		84.8		3.96		27.1	
	SB	4.93	4.19	94.0	9.2	3.23	−0.73	6.1	−21.0
RBR	MB	0.86		88.7		3.40		19.7	
	SB	2.86	2.00	91.3	2.6	2.85	−0.54	5.9	−13.8
SCG	MB	0.86		79.1		4.70		26.6	
	SB	2.75	1.89	89.1	10.0	3.17	−1.53	4.8	−21.8
UGE	MB	0.97		86.7		2.86		22.2	
	SB	2.82	1.85	88.7	2.0	6.21	3.35	8.8	−13.4
RLS	MB	1.19		70.1		4.33		27.6	
	SB	4.39	3.20	76.1	6.0	2.55	−1.78	7.0	−20.6
ZGR	MB	0.62		84.2		3.64		24.8	
	SB	7.57	6.95	88.0	3.8	4.86	1.22	11.5	−13.3
JCO	MB	1.27		73.7		3.21		32.1	
	SB	4.67	3.40	77.3	3.6	1.93	−1.29	8.8	−23.3
mean MB		0.96		81.2		3.7		26.8	
SD MB		0.23		6.5		0.6		4.7	
mean Δ			3.16		6.2		−0.15		−19.4

Appendix 9: Table 3.6. Mean SEM of the breath-by-breath indices

mode	F_{CO_2et} vol %	PIE ml	$slope_{CO_2}$ vol %/l	alv.eff.-CO_2 %
CPPV	0.02	2.5	0.05	0.18
IMV-MB	0.11	5.1	0.09	1.24
IMV-SB	0.08	3.6	0.87	1.24
CPAP	0.04	2.4	0.20	0.59
total	0.07	3.6	0.45	0.94

References

Abdul-Rasool IH, Chamberlain JH (1983) Respiratory gas exchange before and after cardiac operations. J Thorac Cardiovasc Surg 85:856–863

Agostoni E (1972) Mechanics of the pleural space. Physiol Rev 52:57–128

Aitken RS, Clarke-Kennedy AC (1928) On the fluctuation in the composition of the alveolar air during the respiratory cycle in exercise. J Physiol (Lond) 65:389–411

Baker AB, Babington PCB, Colliss JE, Cowie RW (1977a) Effects of varying inspiratory flow waveform and time in IPPV I: Introduction and methods. Br J Anaesth 49:1207–1219

Baker AB, Colliss JE, Cowie RW (1977b) Effects of varying inspiratory flow waveform and time in IPPV II: Various physiological variables. Br J Anaesth 49:1221–1233

Basch von SSK (1887) Über die Funktion des Capillardruckes in den Lungenalveolen, Wien Med Bl 15:465–467

Becker E (1977) Technische Strömungslehre. Teubner, Stuttgart, p 66

Bendixen HH, Hedley-Whyte J, Laver MB (1963) Impaired oxygenation in surgical patients during general anesthesia with control led ventilation. N Engl J Med 269:991

Bergman NA (1969) Effect of varying respiratory waveforms on distribution of inspired gas during artificial ventilation. Am Rev Respir Dis 100:518–525

Brunner J, Langenstein H, Wolff G (1983) Die Unmöglichkeit der direkten Flußmessung für Lungenfunktionsuntersuchungen. Schweiz Med Wochenschr 113:1130–1133

Brunner J (1985) Computerunterstützte, nichtinvasive Methoden zur Analyse von Lungenfunktionen beatmeter Patienten: Untersuchungen zur Optimierung der Beatmung. Diss. ETH. Nr. 7892, Zürich

Brunner J, Wolff G (1985a) Reliable estimation of series dead space in ventilated patients. Clin Physiol 5 (Suppl 3): 65–68

Brunner J, Wolff G, Langenstein H, Cumming G (1985b) Reliable detection of inspiration and expiration by computer. Int J Clin Monit Comput 1:221–226

Brunner J, Wolff G, Cumming G, Langenstein H (1985c) Accurate measurement of N_2 washout requires dynamic adjustment of delay time. J Appl Physiol 59:1008–1012

Campbell EJ, Nunn JF, Peckett BW (1958) A comparison of artificial ventilation and spontaneous respiration with particular reference to V/Q relationships. Br J Anaesth 30:166–175

Chapman S, Cowling TG (1960) The mathematical theory of non-uniform gases. Cambridge University Press, Cambridge

Cheney FW Jr, Burnham SC (1971) Effect of ventilatory pattern on oxygenation in pulmonary edema. J Appl Physiol 31:909–912

Connors AF Jr, McCaffree DR, Gray BA (1981) Effect of inspiratory flow rate on gas exchange during mechanical ventilation. Am Rev Respir Dis 124:537–543

Connors AF Jr, Castele RJ, Altose MD (1983) The effect of inspiratory flow rate on lung volume and distribution of ventilation. Am Rev Respir Dis 4:229

Cooper EA (1961) On respiratory dead space and the adequacy of artificial ventilation. Thesis, University of Cambridge

Craven N, Sidwall G, West P, McCarthy S, Cherniack RM (1976) Computer analysis of the single-breath nitrogen washout curve. Am Rev Respir Dis 113:445–449

Cullen P, Modell JH, Kirby RR, Klein EF, Long W (1975) Treatment of patients with flail chest by intermittent mandatory ventilation and PEEP. Crit Care Med 3:45

Cumming G (1984) Is closing volume a closed volume? Intensive Care Med 10:163–164
Cumming G, Semple SJ (1977) Disorders of the respiratory system, 2nd edn. Blackwell, London
Cumming G, Abraham AS, Horsfield K, Prowse K (1970) Regional hypoxia and dependent airway closure. Scand J Respir Dis 51:37–41
D'Angelo F, Sant'Ambrogio KG, Agostoni E (1974) Effect of diaphragm activity or paralysis on distribution of pleural pressure. J Appl Physiol 37:311
Davies GG, Jebson PJR, Glasgow BM, Hess DR (1986) Continuous Fick cardiac output compared to thermodilution cardiac output. Crit Care Med 14:881–885
Douglas ME, Downs JB (1980) Cardiopulmonary effects of intermittent mandatory ventilation. Int Anesthesiol Clin 2:97–121
Downs JB (1980) Inappropriate applications of IMV. Chest 78:897
Downs JB, Mitchel LA (1976) Pulmonary effects of ventilatory pattern following cardiopulmonary bypass. Crit Care Med 4:295
Downs JB, Klein EF, Desautels D, Modell JH, Kirby RR (1973) Intermittent mandatory ventilation: A new approach to weaning patients from mechanical ventilators. Chest 64:331–335
Drummond GB, Wright ADG (1983) Inaccuracy of oesophageal pressure for pleural pressure estimation in supine anaesthetized subjects. Br J Anaesth 55:585–593
Engel LA, Grassino A, Anthonisen NR (1975) Demonstration of airway closure in man. J Appl Physiol 38:1117–1125
Fairlay HB, Blenkarn GD (1966) Effect of pulmonary gas exchange on variations in inspiratory flow rate during intermittent positive pressure ventilation. Br J Anaesth 38:320
Falke KJ, Pontoppidan H, Kumar A, Leith DE, Geffin B, Laver MB (1972) Ventilation with end-expiratory pressure in acute lung disease. J Clin Invest 51:2315–2323
Farhi LW, Yokoyama T (1967) Effects of ventilation-perfusion inequality on elimination of inert gases. Respir Physiol 3:12–20
Felton CR, Montenegro HD, Saidel GM (1984) Inspiratory flow effects on mechanically ventilated patients: Lung volume, inhomogeneity and arterial oxygenation. Intensive Care Med 10:281–286
Fenn WO, Rahn H, Otis AB (1946) A theoretical study of the composition of the alveolar air at altitude. Am J Physiol 146:637–653
Fenn WO, Otis AB, Rahn H, Chadwick LE, Hegnauer AH (1947) Displacement of blood from the lungs by pressure breathing. Am J Physiol 151:258
Finlay WEI, Wightman AE, Adams AP, Sykes MK (1970) The effect of variations in I:E ratio on cardiorespiratory function during controlled ventilation in normo, hypo- and hypervolaemia. Br J Anaesth 42:935
Fleisch A (1925) Der Pneumotachograph: Ein Apparat zur Geschwindigkeitsregistrierung der Atemluft. Arch Ges Physiol 209:713–722
Fletcher R (1980) The single breath test for carbon dioxide. Thesis, Berlings Arloev Lund, Sweden
Fletcher R, Jonson B, Cumming G, Brew J (1981) The concept of deadspace with special reference to the single breath test for carbon dioxide. Br J Anaesth 53:77–88
Fowler WS (1948) Lung function studies II: The respiratory dead space. Am J Physiol 154:405–416
Francis G, Gelfand R, Peterson RE (1979) Effects of gas density on the frequency response of gas filled pressure transducer. J Appl Physiol 47:631–637
Fredberg JJ, Wohl ME, Glass GM, Dorkin HL (1980) Airway area by acoustic reflections measured at the mouth. J Appl Physiol 48:749–758
Froese AD, Bryan AC (1974) Effects of anaesthesia and paralysis on diaphragmatic mechanisms in man. Anaesthesiology 41:242
Fukuchi Y, Cosio M, Kelly S, Engel LA (1977) Influence of pericardial fluid on cardiogenic gas mixing in the lung. J Appl Physiol 42:5–12
Fuleihan SF, Wilson RS, Pontoppidan H (1976) Effect of mechanical ventilation with end-inspiratory pause on blood-gas exchange. Anesth Analg 55:122–130
Gehan EA, George SL (1970) Estimation of human body surface area from height and weight. Cancer Chemother Rep 54:225–235
Geiger K, Wolff G (1973) Der Einfluß eines verlängerten Inspiriums bei Beatmung wegen respiratorischer Insuffizienz. Thoraxchirurgie 21:414–417
Gieck K (1981) Technische Formelsammlung. Gieck, Heilbronn
Gray BA, McCaffree DR, Sivak ED, McCurdy HT (1978) Effect of pulmonary vascular engorgement on respiratory mechanics in the dog. J Appl Physiol 45:119–127

Grimby G, Hedenstierna G , Loefstroem B (1975) Chest wall mechanics during artificial ventilation. J Appl Physiol 38:576

Hallbaeck I, Karlsson E, Ekblom B (1977) The MGA 200 mass spectrometer as analyzer of the CO_2 and O_2 concentrations in air gas mixtures. Centron Med Appl Bull 013-6

Hansen JE, Ampaya EP, Bryant GH, Navin JJ (1975) Branching pattern of airways and air spaces of a single human terminal bronchiole. J Appl Physiol 38:983–989

Hart MC, Orzalesi MM, Cook CD (1963) Relation between anatomic respiratory dead space and body size and lung volume. J. Appl Physiol. 18:519–522

Hastings PR, Bushnell LS, Skillman JJ, Weintraub RM, Hedley-Whyte J (1980) Cardiorespiratory dynamics during weaning with IMV versus spontaneous ventilation in good-risk cardiac-surgery patients. Anesthesiology 53:429–431

Hauge A, Bo G, Waaler BA (1975) Interrelations between pulmonary liquid volumes and lung compliance. J Appl Physiol 38:608–614

Hedenstierna G, Lundberg S (1975 a) Airway compliance during artificial ventilation. Br J Anaesth 47:1277–1281

Hedenstierna G, McCarthy G (1975 b) The effect of anaesthesia and IPPV with different frequencies on the anatomical and alveolar dead space. Br J Anaesth 47:847–852

Hedenstierna G (1987) personal communication

Hewlett AM, Hulands GH, Nunn JF, Heath JR (1974) Functional Residual Capacity during Anaesthesia, II: Spontaneous Respiration. Br J Anaesth 46:486–494

Hewlett AM, Hulands GH, Nunn JF, Milledge JS (1974) Functional Residual Capacity during Anaesthesia, III: Artificial Ventilation. Br J Anaesth 46:495–503

Higgs BD, Behrakis PK, Bevan DR, Milic-Emili J (1983) Measurement of pleural pressure with esophageal balloon in anesthetized humans. Anesthesiology 59:340–343

Hlastala MP, Scheid P, Piiper J (1981) Interpretation of inert gas retention and excretion in the presence of stratified inhomogeneity. Respir Physiol 46:247–259

Hobelmann CF, Smith DE, Virgilio RW, Peters RM (1977) Mechanics of ventilation with positive end-expiratory pressure. Ann Thorac Surg 24:68–76

Hoppin FG, Hildebrandt J (1977) Mechanical properties of the lung. In: West JB (ed) Bioengineering aspects of the lung, vol 2. Marcel Dekker, New York, pp 83–162

Horsfield K, Cumming G (1968) Functional consequences of airway morphology. J Appl Physiol 24:384–390

Jackson AC, Vinegar A (1979) A technique for measuring frequency response of pressure and flow transducers. J Appl Physiol 47:462–467

James AT, Martin AJP (1952) Gas-liquid partition chromatography: The separation and micro estimation of volatile fatty acids from formic acid to dodecanoic acid. Biochem J 50:679–690

Jansen JRC, Schreuder JJ, Bogaard JM, van Rooyen W, Versprille A (1981) Thermodilution technique for measurement of cardiac output. J Appl Physiol 50:584–591

Jindal SK, Lakshminarayan S, Kirk W, Butler J (1984) Acute increase in anastomotic bronchial blood flow after pulmonary arterial obstruction. J Appl Physiol 57:424–428

Katz JA, Ozanne GM, Zinn SE, Fairlay HB (1981) Time course and mechanisms of lung-volume increase with PEEP in acute pulmonary failure. Anesthesiology 54:9–16

Kelman GR (1966) Digital computer subroutine for the conversion of oxygen tension into saturation. J Appl Physiol 21:1375–1376

Kelman GR (1967) Digital computer procedure for the conversion of P_{CO_2} into blood CO_2 content. Respir Physiol 3:111–115

Knelson JH, Howatt WF, De Muth GR (1970) Effect of respiratory pattern on alveolar gas exchange. J Appl Physiol 29:328

Kohlrausch F (1968) Praktische Physik. Teubner, Stuttgart

Kox W, Langley F, Horsfield K, Cumming G (1982) Effect of atropine on alveolar gas mixing in man. Clinical Science 62:549–551

Kuckelt W , Scharfenberg J, Mrochen H, Dauberschmidt R, Petrakov G, Kassil W, Meyer M (1981) Effect of PEEP on gas exchange, pulmonary mechanics, and hemodynamics in adult respiratory distress syndrome (ARDS). Intensive Care Med 7:177–185

Kumar A, Falke KJ, Geffin B, Aldredge CF, Laver MB, Loewenstein E, Pontoppidan H (1970) Continous positive-pressure ventilation in acute respiratory failure. N Engl J Med 283:1430–1436

LaForce RC, Lewis MB (1970) Diffusional transport in the human lung. J Appl Physiol 28:291–298

Lassen HCA (1953) A preliminary report on the 1952 epidemic of poliomyelitis in Copenhagen with special reference to the treatment of acute respiratory insufficiency. Lancet 1:37
Lemaire F, Jardin F, Regnier B, Loisance D, Goudot B, Lange F, Eveleigh MC, Teisseire B, Laurent D, Rapin M (1978) Pulmonary gas exchange during venoarterial bypass with a membrane lung for acute respiratory failure. Thorac Cardiovasc Surg 75:839–846
Lemen RJ, Gerdes CB, Wegmann MJ, Perrin KJ (1982) Frequency spectra of flow and volume events for forced vital capacity manueouvres. J Appl Physiol 53:977–984
Lerou JGC (1984) The use of long sampling tubes in respiratory mass spectrometry. Dissertation, Institute of Anaesthesiology, Catholic University of Nijmegen
Luce JM, Pierson DJ, Hudson LD (1981) Intermittent mandatory ventilation. Chest 79:678–685
Lyager S (1968) Influence of flow pattern on the distribution of respiratory air during intermittent positive pressure ventilation. Acta Anaesthesiol Scand 12:191–211
Lyager S (1970) Ventilation/perfusion ratio during intermittent positive-pressure ventilation. Acta Anaesthesiol Scand 14:211–232
McCall CB, Hyatt RE, Nobel FW, Fry DL (1957) Harmonic content of certain respiratory flow phenomena of normal individuals. J Appl Physiol 10:215–218
McCarthy DS, Spencer R, Greene R, Milic-Emili J (1972) Measurement of "closing volume" as a simple and sensitive test for early detection of small airway disease. Am J Med 52:747–753
Martin CJ, Das S, Young AC (1979) Measurement of the dead space volume. J Appl Physiol 47:319–324
Milic-Emili J, Mead J, Turner JM, Glauser EM (1964) Improved technique for estimating pleural pressure from esophageal balloons. J Appl Physiol 19:207–211
Milic-Emili J, Henderson JAM, Dolovich MB, Trop D, Kaneko K (1966) Regional distribution of inspired gas in the lung. J Appl Physiol 21:749–759
Nosbaum J, Baer E, Wolff G (1974) Der Einfluß des Atemzugsvolumens auf den intrapulmonalen Rechts-links-shunt, die Totraumventilation und die Hämodynamik bei mechanisch Beatmeten. Schweiz Med Wochenschr 104:1516–1521
Nunn JF (1977) Applied respiratory physiology, 2nd edn. Butterworths, London
Nunn JF, Hill DW (1960) Respiratory dead space and arterial to end-tidal CO_2 tension difference in anesthetized man. J Appl Physiol 15:383–389
Nyffeler W, Brunner J, Wolff G (1984) Easy adjustment modification of monoghan M-250/ohio M-560 ventilator for facile use with IMV, CPPV and nitrogen washout. Acute Care 10:43–46
Paiva M (1972) Computation of the boundary conditions for diffusion in human lung. Comput Biomed Res 5:585–595
Paiva M, Lacquet LM (1976) Gas transport in a model from Hansen-Ampaya data of human lung. J Appl Physiol 41:115–119
Perez-Chada RD, Gardaz J-P, Madgwick RG, Sykes MK (1983) Cardiorespiratory effects of an inspiratory hold and continuous positive pressure ventilation in goats. Intensive Care Med 9:263–269
Piiper J, Koepchen HP (1975) Atmung, 2nd edn. Urban and Schwarzenberg, Munich
Pinsky MR (1984) Determination of pulmonary arterial flow variation during respiration. J Appl Physiol 56:1237–1245
Prakash O, Meij SH, van der Borden SG (1982) Spontaneous ventilation test vs. intermittent mandatory ventilation. Chest 81:403
Prakash O, Meij SH (1985) Oxygen consumption and blood gas exchange during controlled and intermittent mandatory ventilation after cardiac surgery. Crit Care Med 13:556–559
Proctor DF (1964) Physiology of the upper airway. In: Handbook of Physiology Sctn 3, Vol 1: Respiration. Washington, DC. Am Physiol Soc
Prowse K, Cumming G (1973) Effects of lung volume and disease on the lung nitrogen decay curve. J Appl Physiol 34:23–33
Puluj J (1879) Wien Sitzungsber 79:97–113, 745–756, 578–592, 633–642
Quanjer PT (1977) Documentation no SLF-20 of the working group on standardization of lung function tests. Commission of the European Community of Coal and Steel, Luxembourg
Rehder K, Sessler AD, Marsh HM (1975) General anesthesia and the lung. Am Rev Respir Dis 112:541–563
Riley RL, Cournand A (1949) "Ideal" alveolar air and the analysis of ventilation-perfusion relationships in the lungs. J Appl Physiol 1:825–847

Rohrer F, Nakasone K, Wirz K (1925) Physiologie der Atembewegung. Springer, Berlin, pp 70–127 (Handbuch der normalen und pathologischen Physiologie vol 2)
Rossier PH, Bühlmann AA, Wiesinger K (1960) Respiration. In: Luchsinger PC, Moser KM (eds) Physiologic principles and their clinical applications. Mosby, St Louis, p 58
Saidel GM, Salmon RB, Chester EH (1975) Moment analysis of multibreath lung washout. J Appl Physiol 38:328–334
Sanije J, Saidel GM, Chester EH (1979) Real-time moment analysis of pulmonary nitrogen washout. J Appl Physiol 46:1184–1190
Schmitt K (1909) Über die innere Reibung einiger Gase und Gasgemische bei verschiedenen Temperaturen. Ann Phys 30:393–410
Severinghaus JW, Swenson EW, Finley TN, Lategola MT, Williams J (1961) Unilateral hypoventilation produced in dogs by occluding one pulmonary artery. J Appl Physiol 16:53–60
Shepard RH, Campbell EJM, Martin HB, Enns T (1957) Factors affecting the pulmonary dead space as determined by the single breath analysis. J Appl Physiol 11:241–244
Shutack JG, Fox WW, Shaffer TH, Schwartz JG, Moomjian AS (1982) Effect of low-rate intermittent mandatory ventilation on pulmonary function of low-birth-weight infants. J Pediatr 100:799–802
Skarvan K, Graedel E, Hasse J, Stulz P, Pfisterer M (1984) Coronary artery spasms after coronary artery bypass surgery. Anesthesiology 61:323–327
Steenblock U, Mannhart H, Wolff G (1976) Effect of hemorrhagic shock on intrapulmonary right-to-left shunt (Q'_S/Q'_T) and dead space (V_D/V_T). Respiration 33:133–142
Strandberg A, Tokics L, Brismar B, Landquist H, Hedenstierna G (1986) Atelectasis during anaesthesia and in the postoperative period. Acta Anaesthesiol Scand 30:1154–1158
Suter PM, Brigljevic J, Hemmer M, Gemperle M (1977) Effets de la pause en fin d'inspiration sur l'échange gazeux et l'hemodynamique chez des patients en ventilation méchanique. Can Anaesth Soc J 24:550–558
Suter PM, Fairlay HB, Isenberg MD (1978) Effect of tidal volume and positive end-expiratory pressure on compliance during mechanical ventilation. Chest 73:158–162
Sykes MK, Lumley J (1969) The effect of varying inspiratory:expiratory ratios on gas exchange during anaesthesia for open – heart surgery. Br J Anaesth 41:374
Uhl RR, Lewis FJ (1974) Digital computer calculation of human pulmonary mechanics using a least squares fit technique. Comput Biomed Res 7:489–495
Wagner PD, Saltzmann HA, West JB (1974) Measurement of continuous distributions of V/Q ratios: theory. J Appl Physiol 36:588–599
Wald A, Jason D, Murphy TW, Mazzia VDB (1969) A computers system for respiratory parameters. Comput Biomed Res 2:411–429
Watson WE (1962a) Some observations on dynamic lung compliance during intermittent positive pressure respiration. Br J Anaesth 34:153
Watson WE (1962b) Observations on physiological deadspace during intermittent positive pressure ventilation. Br J Anaesth 34:502
Weast RC (ed) Handbook of Chemistry and Physics (1974–75) CRC Press, Cleveland Ohio
Weibel ER (1963) Morphometry of the human lung. Springer, Berlin Göttingen Heidelberg
Weisman IM, Rinaldo JE, Rogers RM, Sanders MH (1983) Intermittent mandatory ventilation. Am Rev Respir Dis 127:641–647
Weiss JW, Rossing TH, Ingram RH (1983) Effect of intermittent mandatory ventilation on respiratory drive and timing. Am Rev Respir Dis 127:705–708
Werner O, Malmkvist G, Beckman A, Stahle S, Nordström L (1984) Gas exchange and haemodynamics during thoracotomy. Br J Anaesth 56:1343–1349
West JB (1969) Ventilation-perfusion inequality and overall gas exchange in computer models of the lung. Respir Physiol 7:88–110
West JB, Fowler KT, Hugh-Jones P, O'Donnell TV (1957) Measurement of the ventilation-perfusion ratio inequality in the lung by the analysis of a single expirate. Clin Sci 16:529–547
Westbrook PR, Stubbs SE, Sessler AD, Rehder K, Hyatt RE (1973) Effects of anesthesia and muscle paralysis on respiratory mechanics in normal man. J Appl Physiol 34:81–86
Westenskow DR, Cutler CA, Wallace WD (1984) Instrumentation for monitoring gas exchange and metabolic rate in critically ill patients. Crit Care Med 12:183–187
Woldenberg MJ, Cumming G, Harding K, Horsfield K, Prowse K, Singhal S (1970) Law and order in the Human Lung. Harvard papers in theoretical Geography Nr 41. Office Naval Research Report (AD 709602) Nat'l Techn Inf Service, US Dept Commerce, Springfield VA.

Wolff G (1983) Die künstliche Beatmung auf Intensivstationen, 3rd edn. Springer, Berlin Heidelberg New York

Wolff G, Brunner JX (1984) Series dead space volume assessed as the mean value of a distribution function. J Clin Monit Comput 1:177–181

Wolff G, Grädel E, Claudi B, Rist M, Schwab T (1972) Der Einfluß des akut erniedrigten Herzminutenvolumens auf den intrapulmonalen Rechts-links-Shunt. Schweiz Med Wochenschr 102:198–202

Wolff G, Langenstein H, Schwendener R, Lischer P (1982) Optimal end-expiratory airway pressure for ventilated patients. Intensive Care Med 8:39–48

Wolff G, Brunner JX, Grädel E (1986) Gas exchange during mechanical ventilation and spontaneous breathing. Intermittent mandatory ventilation after open heart surgery. Chest 89:11–17

Yeh MP, Gardner RM, Adams TD, Yanowitz FG (1982) Computerized determination of pneumotachometer characteristics using a calibrated syringe. J Appl Physiol 53:280–285

Subject Index

GPSR Compliance

The European Union's (EU) General Product Safety Regulation (GPSR) is a set of rules that requires consumer products to be safe and our obligations to ensure this.

If you have any concerns about our products, you can contact us on ProductSafety@springernature.com

In case Publisher is established outside the EU, the EU authorized representative is:

Springer Nature Customer Service Center GmbH
Europaplatz 3
69115 Heidelberg, Germany

Batch number: 10704771

Printed by Printforce, the Netherlands